D1107443

DISCARD

Praise for
Global Warming I$ Good for Business

"Contrary to popular belief, successful entrepreneurs rarely start a business to create a new mousetrap. Research supports that identifying a pain and then creating a solution to cure it is the root cause of most successful ventures.

Never before has there been the collection of pain that there is in our environment. Ms. Keilbach shows us every tangent of that pain is an opportunity waiting to be implemented. Global warming and its consequences will be cured by that misunderstood collection of society known as entrepreneurs."

—Tom O'Malia, director, Lloyd Greif Center for Entrepreneurial Studies, Marshall School of Business, University of Southern California

"An environmentalist sees an environmental problem and says, 'Let's regulate it!' An entrepreneur sees an environmental problem and says, 'How can I make money by solving it?' Clearly, there's a world of difference in these two approaches, as Kimberly Keilbach makes eminently clear in her useful and timely new book, *Global Warming I$ Good For Business*. All those seeking politically palatable strategies for saving the planet should read this book."

—Byron Kennard, executive director, The Center for Small Business and the Environment

"Some leading companies are beginning to grasp the true meaning of the title of Kimberly Keilbach's shrewd and insightful new book and are seizing the many business opportunities offered by a carbon-constrained future. These will be the truly successful companies of the 21st Century."

—Lloyd Timberlake, director, North America office, World Business Council for Sustainable Development

Global Warming I$ Good for Business

How Savvy Entrepreneuers, Large Corporations,
and Others Are Making Money While Saving the Planet

K.B. Keilbach

Fresno, California

Printed in the United States of America.

Published by Quill Driver Books
an imprint of Linden Publishing, Inc.
2006 South Mary, Fresno, CA 93721
559-233-6633 / 800-345-4447
QuillDriverBooks.com

Quill Driver Books may be purchased for educational, fund-raising, business or promotional use. Please contact Special Markets, Quill Driver Books at the above address or phone numbers.

Quill Driver Books Project Cadre:
Doris Hall, Christine Hernandez, John David Marion,
Stephen Blake Mettee, Kent Sorsky, Maura J. Zimmer

135798642

ISBN 978-1884956-88-1 • 1-884956-88-2

To order a copy of this book, please call 1-800-345-4447.

Library of Congress Cataloging-in-Publication Data

Keilbach, K. B. (Kimberly B.), 1960-
  Global warming is good for business : how savvy entrepreneurs, large cor-
porations, and others are making money while saving the planet / by K.B.
Keilbach.
    p. cm.
  ISBN-13: 978-1-884956-88-1
  ISBN-10: 1-884956-88-2
1. Global warming—Economic aspects—United States. 2. Climatic
changes—Economic aspects—United States. I. Title.
  QC981.8.G56K45 2008
  363.738'74—dc22   2008033864

For Mark, Paul & Dan—
May you always see the opportunity in change

Contents

Acknowledgments

I would like to thank my publisher, Steve Mettee, who first saw the possibilities of this book, as well as my agent, Angela Rinaldi, whose guidance has been invaluable in the writing process. I would also like to thank Shelly Lowenkopf, Noel Riley Fitch, and Paula Brancato, from whose classes this project developed.

Karen Eschenbach, Erik Silverberg, Ed and Pauline Brower all dedicated a substantial amount of time and energy to read over and critique my manuscript, and I greatly appreciate their help, encouragement and advice. I also appreciate the support that my husband, Roger, has given me to finish this project.

Lastly, I would like to thank all of those who took time from their hectic routines to patiently answer my questions and offer useful information. Your thoughts and comments, whether directly used here or not, have been both thought-provoking and insightful.

Introduction
What part of climate change don't you understand?

You could say this book began with a cup of coffee. I was sitting at a Starbuck's waiting for a friend one day when I read "the way I see it" on the back of my cup. It was a quote by Chip Giller, founder of Grist.org, which said: "So called 'global warming' is just a secret ploy by wacko tree-huggers to make America energy independent, clean our air and water, improve the fuel efficiency of our vehicles, kick-start twenty-first-century industries, and make our cities safer and more livable. Don't let them get away with it!"

I had to laugh. "Imagine that," I thought. "Imagine jumping on board the global warming bandwagon *because* you want a stronger economy, not in spite of it. Imagine climate change as the impetus for a shift in culture. Imagine the energy crisis fueling innovation and a better way of life." I began to look at the positive aspects of climate change and energy shortages, and I found something very interesting.

This is not the first energy or environmental crisis the human race has ever faced. Preindustrial Europe was plagued by fuel shortages which burdened the poor and middle classes and threatened the economy. The energy crisis at that time stemmed from centuries of deforestation and a universal shortage of wood. The alternative fuel of the day, a low-grade coal, produced so much soot and smog, especially in the cities, that it created a health hazard for those unlucky enough to live there. Individuals sued their industrial neighbors in an attempt to stem the tide of airborne toxins, as government regulations fell abysmally short of protecting its citizens. Still, it was better than nothing, and, as so often happens, the economic needs of society trumped the environment, as well as concerns for the overall health of the population.

One of the first innovative solutions to this crisis came in 1603, when a man named Sir Henry Platt came up with the idea of charring coal to remove its impurities. The resulting coal derivative, known as coke, burned hotter and cleaner than either coal or charcoal, so much so that by the middle of the 1600s, coke was used commercially to roast malt for beer, one of the industries in England at the time.

Then, in 1709, Abraham Darby came up with another innovative idea: use coke to fuel the forges that produced iron ore. Coke proved to be such an efficient and cost-effective fuel in iron smelting that it became universally used for that and other manufacturing needs. This in turn led to a surge in overall industrial output, which helped to usher in the Industrial Age. The rest, as they say, is history.

To say that these innovations solved all of the energy and environmental problems of the time is a gross oversimplification; there were obviously several complex factors involved, some of which we are still dealing with today. The point here is that, in at least one instance, an energy and environmental crisis has spurred human beings to come up with innovative ideas and new opportunities to improve the human condition. Today, climate change and energy security are two very real concerns, but they also offer us tremendous opportunities for improving our lives and "saving" our planet.

Whether people believe global warming is the direct result of greenhouse gases heating up the surface of the earth or whether they believe we are going through yet another in a long history of cyclical climate changes, virtually everyone agrees that there is *something* happening to our planet and humans do play a part. But what does this mean, really? Many of us in the United States feel helplessly caught up in the debate between what is and is not true and what action (if any) we should take—so much so that, at the end of the day, we do nothing.

Introduction

However, global warming is more than a source of debate; it is a catalyst for positive change—the raw material, if you will, for innovation. Change is uncomfortable, painful even, but it is also motivating. It encourages us to look at the same things we have always done and to find, in them, new ways to progress and move forward. It inspires businesses to find genuinely game-changing technologies that will render the status quo irrelevant, and it compels individuals to grow and develop in order to achieve a higher quality of life.

There is no doubt that the issues we face with regards to climate change are complex and seemingly insurmountable. For every solution, it seems, there is a trade-off, a downside, and it is difficult, if not impossible, to see which will be most beneficial in the long run. Virtually everyone in the world today is looking for effective (and cost-effective) alternatives. Some say the United States is behind on this learning curve. Others look to us to set an example for prosperous, sustainable living.

The idea of being leaders in a global problem-solving process is daunting. But there are a number of people in this country who do see the opportunities for innovation and growth in the environmental and energy crises that face us and who are taking a leadership role in dealing with these issues. Business and government leaders, scientists and engineers, entrepreneurs and venture capitalists, and others are all working towards a solution that could, ultimately, benefit us all.

In fact, there are many individuals and organizations operating in the United States today who have embraced the changes that global warming and energy security have brought about and who are discovering ways to impact our world in a positive manner. They are working toward a greener future, where human beings can build environmentally sustainable lives *and* make a buck.

Admittedly, not every solution they come up with will work, and those that do work technically may not ever reach their market potential. Even the best innovations could take generations to evolve to a point where they actually solve a problem at an affordable price. It is virtually impossible to say which ideas are here to stay and which are not. Doubtless, some of the hottest new inventions today will go the way of the buggy whip. Others will morph into something else entirely. But, maybe, just maybe, one of them will be the tipping point for a whole new age.

In the meantime, whether they are motivated by a genuine concern for the environment, intrigued by new discoveries in research and development, spurred by government incentives, or simply in it for their own economic benefit, people are finding it is more profitable to implement sustainable environmental practices and develop "green" products and "cleantech" services than ever before.

Bottom line: Global warming *is* good for business.

Part One
The Incubators

1

Stabilization Wedges

On October 12, 2007, students from universities all over the world converged on the National Mall, a strip of green grass and gravel walkways that runs from the United States Capitol to the Lincoln Memorial. Some had traveled for days to get there. Many had spent as much as two years researching, designing, and constructing their projects. All shared a common concern for sustainability and the environment.

Twenty solar houses, no more than 800 square feet each in area, stood neatly in rows along the mall, while those who had designed and then built them stood expectantly beside them, waiting to share the fruits of their labors. The 3rd Annual Solar Decathlon had begun.

Sponsored by a broad range of government agencies, associations, and businesses, the universities who competed in the 2007 decathlon brought cutting-edge science and technology quite literally to the street. The interdisciplinary teams consisted of students from a variety of disciplines ranging from philosophy to thermal engineering, each with their own diverse set of skills and interests.

Projects were judged and points given in ten areas: architecture, engineering, market viability, communications (each team offered public tours), interior comfort, appliances, ability to deliver hot water, lighting, overall energy balance, and "getting around." At the end of the competition, extra energy generated by each house's solar system was used to fuel each team's electric vehicles, and points were given to the team who could drive the farthest.

Global Warming I$ Good for Business

There was a carnival-like atmosphere at the decathlon during that sunny week. Total strangers talked together like old friends, exchanging ideas and information about various technologies. They learned about passive, photovoltaic (PV) and thermal solar systems (and combinations of systems), about battery storage types and capacities, and about basic building structure and design. It was an amazing experience—alternately astonishing, confusing, exhilarating, and exhausting.

Each house was unique in design, function, and use of renewable building materials. Some looked like they came from a futuristic movie set. Others looked downright homey. But all of them were more than just houses—they were a glimpse into what the future might look like. And it was pretty darn good!

In the end, when all the points were tallied and the results were announced, the house from Technische Universitat Darmstadt (Germany) came in first, with the University of Maryland's LEAFHouse coming in second and Santa Clara University's Ripple Home coming in third. But the event was really a win for everyone interested in renewable energy. It drew over 100,000 people, many of whom had little or no prior understanding of solar energy. And it got a lot of them to think about the real-life possibilities of sustainable living.

It seems like everyone today is interested in global warming, but a lot of people cannot tell you what global warming actually is. The short version is that global warming has to do with an increase in the Earth's temperature due to what is commonly known as the greenhouse effect.

Anyone who has ever sat in a closed car on a hot summer day has experienced the greenhouse effect. Simply put, energy from the sun heats up the air in the car. The air becomes trapped behind the windshield and continues to get progressively—and often lethally—hotter. Many scientists believe this is what is going on right now on Earth. According to them, the heat of the sun is being trapped behind greenhouse gases in much the same way as it might be trapped behind the windshield of a car.

Another term frequently used in environmental discussion is climate change. Climate change represents an overall change in long-term weather patterns and is not specific to temperature. For example, climate change may refer to an increase or a decrease in rainfall over time. In 2005, the National Academies reported, "The phrase 'climate change' is growing in preferred use to 'global warming' because it helps convey that there are changes in addition to rising temperatures."

Our climate has been changing since the Earth first formed. The question now is how much of the change is due to human-caused greenhouse gas emissions and what can we do to reduce our emissions.

The major greenhouse gases include carbon dioxide (CO_2), methane (CH_4), nitrous oxide (N_2O), and fluorinated gases (HFCs, PFCs and SF_6). Of these, according to the National Energy Information Center, carbon dioxide emissions represent over 80 percent of total human-made greenhouse gas emissions in the United States as of 2007.

It is important to note that greenhouse gases and smog are not the same thing. Smog is a kind of air pollution—the product of a photochemical reaction between sunlight, hydrocarbons, and nitrogen oxides. Currently, the United States Environmental Protection Agency has established National Ambient Air Quality Standards (NAAQS) to measure six air contaminants: carbon monoxide (CO), lead (Pb), nitrogen dioxide (NO_2), ozone (O_3), particulate matter (PM), and sulfur dioxide (SO_2). However, reducing those contaminants—while it would improve our air quality—would not necessarily reduce the so-called "greenhouse effect." The question is, what would?

In 2004, physicist Robert Socolow and ecologist Stephen Pacala of Princeton University developed the concept of "stabilization wedges [for] solving the climate problem for the next fifty years with current technologies." These included becoming more energy efficient and conserving energy; shifting our source of fuel (i.e. from coal to natural gas); capturing and storing carbon dioxide (CO_2);

developing nuclear fission; developing renewable electricity and fuels; and implementing sustainable forestry and agricultural soil practices. Although no one of these current technologies can completely stabilize the rapidly climbing business-as-usual trajectory for global carbon emissions, Pacala and Socolow suggested a combination of them together could flatten the ramp of that trajectory to a more sustainable level.

Currently, major universities in the United States are devoting significant resources to studying one or more of these "potential wedges" for CO_2 reduction. Many are focusing on the development of renewable energies, such as solar, wind, biomass, and hydroelectric, all of which are naturally occurring and theoretically inexhaustible. Others are looking into alternative energy sources, such as nuclear power, which are not technically renewable but are alternatives to current fossil-based technologies. Still others are looking at increasing energy efficiency—what Duke Energy's CEO Jim Rogers calls "the fifth fuel"—through implementing different modes of transportation, urban design, and building technologies.

In some respects, universities today do much of the work that corporate research and development departments did in the past. Though not always true "incubators" in the strict business sense, they are places where ideas may be hatched and grown to a point where they become viable in the marketplace. Yet university projects often require years and intensive amounts of capital to develop nascent ideas into potentially viable concepts. Many of them never make it any further than the drawing board. Most academics will admit their work is just the beginning—the seed for new technologies that might flourish if the social conditions and economic environments are right.

The technologies that are being explored are as varied as the individuals involved. More and more scientists and researchers are discovering that they must be just as innovative in getting their ideas out to the public as they are in developing those ideas in the first place. Some are being funded by government grants. Others are choosing to partner with corporations both for funding and to maximize the marketing potential of their innovations. Still others are going the venture capital route and are spinning off to develop their own green businesses. Regardless of the technology, it is becoming increasingly evident that any idea born in a lab must eventually cross through what has been termed a "valley of death" to become commercially viable.

The LEAFHouse—Solar Power

A leaf is one of the most elegant mechanisms known to man for capturing the energy of the sun. It is also the acronym for the University of Maryland's LEAFHouse, a working model of how that same energy capture might be accomplished from an architectural and engineering point of view.

Amy Gardner has been an architect for more than twenty years and is an associate professor at the University of Maryland's School of Architecture, Planning, and Preservation. In the fall of 2007, Professor Gardner was the faculty advisor for the university's LEAF (Leading Everyone to an Abundant Future) House, which took second place at the 2007 Solar Decathlon. "No amount of training could prepare anybody for this," she says. "It's an enormous team....There are graduate students, undergraduate students, and we also had a varied mentor network."

In fact, dozens of corporate and organizational sponsors, such as Sanyo Electric (USA) Corporation and the Maryland Energy Administration, donated money, material, and in some instances expertise to the project. In the end, the students—most of whom had never designed, let alone, built anything—put together a simple-yet-elegant zero-energy home based on the concept of a leaf.

A zero-energy home is one that is designed and constructed to produce as much energy as it consumes annually. The United States Department of Energy (DOE) states, "A Zero Energy Home (ZEH) combines state-of-the-art, energy-efficient construction and appliances with commercially available renewable energy systems." All of the homes at the DOE-sponsored Solar Decathlon were designed to be ZEH.

From the outside, the 800-square-foot LEAFHouse looked contemporary in design, with a series of shuttered doors that opened onto a wooden porch and the environment beyond. It was both energy and water efficient and free of toxic materials. The front of the house had what is known as a green wall, a planted vertical garden, which doubled as a greywater filtration and irrigation system. The greywater from the sinks, shower, and washing machine could be filtered and used to maintain the yard or wash a car, thus decreasing the amount of freshwater consumed. Inside, open sunroofs allowed passive lighting into the main room, which had a warm and inviting appeal, belying the state-of-the-art technology that existed behind the walls.

The LEAFHouse featured Smart-House Adaptive Controls, a unique computerized system that monitored energy production according to internal factors, such as personal preference, and external factors, such as weather, in order to maximize the energy efficiency of the home. It also utilized a liquid desiccant waterfall on the living room wall that combined art with function, capturing moisture from the air and thus improving the efficiency of the air-conditioning system. Obviously, everything in the house was run on solar power.

Solar energy has been around since the beginning of the solar system. Human beings have used passive solar energy for centuries as a source of light, to warm our homes and even to dry our clothes. According to LEAFHouse literature, "The earth receives more energy from the sun in *one hour* than the world uses in a whole year."

The question is, how do we harvest that energy and, perhaps even more importantly, where can we store it?

Photovoltaic (PV) systems are one of the most common means of actively harnessing solar energy and converting it into electricity. Light-to-electricity conversion—known as the photovoltaic effect—was first discovered in the late 1830s by French physicist Alexandre Edmond Becquerel.

In 1954, researchers at Bell Laboratories developed the first silicon PV cell. By the 1960s, photovoltaics were used to power systems for space satellites. In 1977, NASA installed PV systems in several meteorological stations. That same year, the Solar Energy Research Institute (later to become NREL) was launched. And in 1978, the first village PV system was installed on the Papago Indian Reservation in Arizona.

Although the basic science of photovoltaics is nothing new, increased energy efficiencies, decreased costs, and new applications or derivations of existing technologies are evolving every day. It was this "ripple effect" of new ideas—and of people spreading the word of those ideas—that Santa Clara University sought to capitalize on in their Ripple Home entry to the 2007 decathlon.

In addition to the recycled-content tile floors, reclaimed-wood furniture, and denim- (as in recycled blue jeans) insulated walls, the home also contained some high-tech features, such as electro-chromatic windows, which lighten or darken depending on how much passive solar energy is desired. For example, in the wintertime, when you want to maximize the amount of passive solar energy in your home, the windows stay clear. In the summertime, the windows darken to provide "shade" from the sun.

Active PV systems (versus passive systems like windows that capture sunlight) traditionally consist of panels mounted to capture the maximum amount of the day's sunlight, usually on a southern-exposure rooftop or similar structure. The silicon cells within the panels absorb the heat of the sun and begin an electrical current flow from the PV module to the home or building via an electrical wiring system.

In the Technische Universitat Darmstadt's entry, the PV modules were mounted on louvers that formed the exterior walls of the house. The louvers also acted as shade for the interior. A tracking system on the home automatically tilted them to follow the sun's path during the day in order to maximize shading and power production. Another feature of the German house was the phase-change materials in the walls and ceiling. Here, the students had embedded "microcapsules of paraffin" that changed from solid to liquid as they were heated and then stored that energy within the interior of the home.

All of the homes in the Solar Decathlon were designed to be off grid, meaning not tied in with any central utility grid. They relied solely on a large bank of batteries to store solar electricity that could last for a period of days if necessary in order to provide power during no or low sun periods.

Most homes and offices use solar electricity that is grid-tied or utility-interactive. In other words, they tie in directly to their local power company's electric grid, just the same as non-solar users. In the grid-tied system there is no battery bank, so it is much more cost-effective. Electricity runs directly to an inverter, which then converts it from DC to AC to power the home. However, in the utility-interactive system, although the user still has the option of connecting to the utility grid, one can also opt to store power in batteries. The obvious benefit for both systems is that they can provide uninterrupted power whenever needed without having to rely on a battery bank.

During the day—which is the peak period for energy usage—PV panels generate electricity. Excess electricity can either be stored in on-site batteries or it can be "banked" with the local utility company by simply turning the meter backwards. This type of program, known as net metering, allows customers to use the excess power they have generated with their on-site power system to offset their consumption over a given billing period.

At night, when the PV panels do not generate electricity, energy can either be retrieved from storage or purchased from the utilities during non-peak hours when the cost of energy is less.

According to Solar Development Inc., a solar energy equipment supplier with over 30 years' experience, "Today, more solar energy is used for heating swimming pools than for any other single use. Over 200,000 pools are heated by solar in the United States alone." That figure does not include pools heated with passive solar (i.e. sunlight).

Solar thermal collectors are another example of active solar energy. Solar thermal collectors consist of a series of evacuated or vacuum sealed tubes, each of which has transfer fluid (i.e. water, antifreeze). The tubes are placed side-by-side to collect heat from the sun's energy. The sun heats up the liquid within the tubes and then the heated liquid is transferred via copper pipe and stored to heat water. It can also be used to heat space or even to "fire up" absorption chillers in order to cool the air within the home.

Over time, the savings from both PV and solar thermal systems may be great; however, the initial cost of installation can be quite expensive, even with government or utility company rebates and incentives. The amount of time for each system to pay for itself often is several years. For this reason, many people have adopted a wait-and-see approach to solar power, meaning basically wait and see if prices go lower as technologies improve.

This kind of logic is lost on people like Amy Gardner who says, "Your car never pays for itself.... It will do nothing but depreciate, and [people] think that's just the way it is. But when you tell someone that to put a solar array on their house is going to cost $20,000 and then it might pay for itself in twenty years, they go, 'Well, that's just not acceptable.'"

For Gardner, it's a matter of thinking differently about how we inhabit the planet. "If we didn't have gas stations today," she says, "and somebody said, 'I have a good idea. We should have gas stations in the cities,' you could never in a million years get anybody to let you do that. People would go, 'Are you serious? Put an underground tank full of gasoline in the middle of the city? That's a terrorist target. It's an environmental disaster waiting to happen. It's polluting.' No one would ever let you do that now."

Gardner admits, "It's kind of a daunting thing for people to contemplate. It's easy to say, 'Everybody has to change.' But then... you think, 'Well, what would I do to make that big change?'" Like many of her colleagues, she sees a consolidation of different energy strategies rather than the prominence of any given one: "I don't think we're going to be consuming any less energy. I think we might be smarter about it, but I don't think we're going to be doing any better." She adds, "It's going to take a long time to turn the *Queen Mary.*"

Yet some headway is being made. Solar Energy Industries Association president Rhone Resch stated during a November 2007 press teleconference, "The photovoltaic or solar electric market is on track to grow by over 60 percent in the United States." And, in a joint statement with the directors of the National Hydropower Association, the Geothermal Energy Association, and the American Wind Association in January 2008, Resch claimed, "Almost 6,000 [megawatts] of new renewable energies came on-line in 2007, infusing over $20 billion of investment into our economy. And along with this investment came...tens of thousands of high quality jobs in all fifty states."

Businesses from Macy's to Microsoft seem to prove his point. They and a variety of organizations across the United States are installing PV solar and other technologies at their facilities. At the same time, the Department of Energy's Solar Energy Technologies Program, working closely "with national laboratories, universities, industry, professional associations, and other programs" is currently researching and developing different types of renewable and alternative energies, including concentrating solar power technologies, such as the trough system, the dish/engine system, and the power tower. Each of these systems utilizes a different configuration of mirrors to convert sunlight into high-temperature heat that is, in turn, transferred to a steam generator to produce large-scale utility grid electricity without producing greenhouse gases.

The future, say proponents of solar power, has never looked so bright.

3

The Power Mix—Wind & Waves

As you drive across the flat, barren Texas Panhandle, you quickly become aware that wind is a constant feature of the landscape. Wind is to West Texas what sunshine is to California, which is why West Texas A & M University is perhaps one of the best places in the world to explore one of the fastest growing forms of renewable energy.

Wind energy production has become big business in the state of Texas, and the number of turbines now surpasses those in any other state in the United States. With hub heights of over 100 meters (330 feet) and blade lengths of over 50 meters (165 feet), some of the newer wind turbines stand as much as forty stories tall from ground to blade tip. The Empire State Building is 102 stories, to give some sense of perspective.

Large utility-scale wind farms can consist of as many as 250 of these wind turbines, each one able to produce 2.5 million watts, or 2.5 megawatts (MW), of electricity per year. One MW of wind "generates about as much electricity as 240–300 households use," according to the American Wind Energy Association (AWEA).

The trick, according to David Carr, is to get the electricity from the wind farm to the city. Carr has both academic and real-world experience. He has studied electronic engineering and technology and has even owned his own electronic repair business. Currently, he works as assistant director of the Alternative Energy Institute at West Texas A & M University. He says, "We're in the best windy area in the state, but there are not a lot of places that use the load here. So we have to find a way to send our power down to Dallas or

Austin or San Antonio or Houston or somewhere where they need a lot of power. And I think that's the same situation in a lot of areas in the United States."

Carr says that one of the projects the state of Texas is working on is redesigning power lines and running new capacity from the southern part of the state to the North Panhandle. Another area of research is in storing the electricity that comes from large wind systems.

"If anybody can figure out the storage situation, then it would totally revolutionize renewable energy because that's the main problem," says Carr. "A lot of times the wind will blow at nighttime, and all the loads are being used in the daytime. If you can't match that up, then it's pretty hard to justify wind to major power companies because they still have to provide that load. They have to keep the lights on."

Carr says, "It's always easier to save energy than to make energy." Even so, he believes that wind is a force to be reckoned with in terms of providing a significant portion of electricity to the overall power grid of the future. He predicts that wind energy will account for as much as 20 percent of the electrical power we consume because "there is nothing that is less environmentally damaging that can do as much good for us, and fossil fuel prices are not going to go down."

The Alternative Energy Institute works closely with the Texas State Energy Conservation Office (SECO), as well as with potential wind farm developers, collecting and analyzing wind data to see where wind farms are most viable. But wind energy is more than giant wind farms. Small wind—usually consisting of one turbine to run a home, farm, or business—is another area of research and development. The quintessential windmills that have been used for generations to pump water for livestock on farms and ranches across the southwest are one example of small wind power.

Modern-day small wind generators, or wind energy converters, usually consist of a 30- to 100-foot tower and three horizontal-axis rotors or blades, which look like a cross between an old-fashioned farm windmill and an airplane propeller. The blades are connected to a drive train that is, in turn, connected to a generator that converts the rotational energy of the blades into electric current. This current runs down the tower and into an inverter which switches the power from DC to AC for home or office use.

Carr says these systems are best utilized on at least an acre or two of property, mainly due to buildings blocking wind and also because of neighbor complaints about the aesthetics of the towers. However, in August 2007, the North Texas town of Grand Prairie announced that it had installed a hybrid solar/wind-powered street light in a "highly visible" part of town. In this case, the wind turbine and solar panel were mounted on top of the thirty-foot light pole near a senior center and the town's main library.

According to Carr, small wind turbines still need further research to become "more rugged" with fewer "maintenance issues" for the average home and business owner: "If we can help bring along that research and get the small wind turbines to the point that the large wind turbines are, then I think we'll see a significant increase in distributed wind in the future."

Distributed wind power comes from small wind installations that are distributed where needed rather than centrally located on a wind farm. Although they may still be connected to the grid, most of the energy from distributed wind installations is used at one facility or business or even a cluster of homes. The ability of small wind turbines to power remote locations, such as outlying farms and ranches and even sailboats and RVs, is one of its attractions.

With wind systems being installed across the country (and one now proposed offshore in the ocean near Cape Cod, Massachusetts), the American Wind Energy Association (AWEA) claims that "development of just 10 percent of the wind potential in the ten windiest...states would...reduce total United States emissions of CO_2 by almost a third."

And, according to AWEA: "The potential of wind to improve the quality of life in the world's developing countries, where more than

2 billion people live with no electricity or prospect of utility service in the foreseeable future, is vast."

While not yet as well developed as wind, wave power is another potential source of renewable energy. The idea of being able to drop a power buoy off the side of a boat and use the motion of the ocean, as it were, to generate electricity is intriguing to many, including the United States Navy, which has entered into a contract with Oregon State University's Wallace Energy Systems and Renewables Facility (WESRF) to research wave power.

Dr. Ted Brekken is codirector of WESRF, where he spends much of his time researching wave power technologies. Brekken studied wind turbine control and wind energy issues as a Fulbright scholar in Norway, and he sees many similarities between the early days of wind energy—which he refers to as a "mature technology"—and wave energy.

For starters, both wind and wave power have utility scale systems as well as smaller scale systems. Utility scale systems can be tied in directly to the grid, while smaller scale systems can be an excellent source of energy in isolated areas, such as remote islands or onboard pleasure craft or even United States Navy ships.

Being a research institution, and not a developer of wave energy, WESRF has formed industrial partnerships with companies such as Columbia Power Technologies and Bonneville Power. It has also received funding through the Small Business Technology Transfer Program, and is working with Peregrine Power as part of that program.

"In the early days of wind energy, there were a lot of designs out there, and the market eventually chose the design that we see today," says Brekken. "But we're not to that point yet with wave energy; there's still lots and lots of different ideas…about how to harvest [power]."

Basically, onshore wave energy systems capture the energy of breaking waves while offshore systems, including those used by seagoing vessels, capture the energy of deep-water waves. Both systems convert wave energy into electricity. Currently, there are no commercial wave farms in the United States; however, "wave parks" are in various stages of deployment in other parts of the world.

One example of an onshore system is the Wavegen Limpet, an oscillating water column, which is a partially submerged structure with an opening below the waterline. The Limpet, located near Islay in Scotland, blends in with the rock formations and looks like a small bunker on the edge of the sea. The air in the column above the waterline is compressed each time the water rises and decompressed each time the water falls, which causes a turbine within the chamber to spin an electrical generator, thus producing electricity.

Offshore systems include point absorber systems, or buoys, which harvest the up-and-down motion of ocean waves and convert it into electricity. According to a December 2007 article in the *Los Angeles Times*, Canadian-based Finavera Renewables has begun discussions with Pacific Gas & Electric to provide eight buoys for energy production along the California coast. The buoys will extend along the surface of the water in groups that transmit electricity via a central cable to the shore.

Other offshore systems, such as the Pelamis Wave Energy Converter, consist of hollow cylinders of steel that float on the surface of the water. The Pelamis, which resembles a giant cylindrical snake, is called an "attenuator system." The motion of the waves causes high-pressure oil within the cylinders to be pumped through hydraulic motors, which drive electrical generators. Currently, multi-unit wave farms are being set up for operation off the coasts of Portugal and Scotland.

The so-called "Danish Wave Dragon" has been tested in Denmark and is what is called an overtopping system that consists of a giant ramp over which waves flow into a reservoir. From there, the captured water flows downward through a turbine, which converts the wave energy into electricity.

According to Brekken, there is nothing "earth shattering" about wave technology. "It's just a matter of applying the technology that exists and doing it in a way that is efficient and survivable in the marine environment. The marine environment is a very, very harsh place to be. It's a very difficult environment to have something that works consistently. So the challenge is trying to make these devices robust and efficient."

Oregon State University has a unique device called the Linear Test Bed (LTB) which mimics the action of waves in a laboratory setting. The LTB tests various wave energy designs without incurring the expense of setting them up in the ocean. One of the challenges currently being studied is how to store the energy that is produced and to "buffer that for the grid." Another area under study is the effects wave farms have on the marine environment.

Brekken says the majority of environmental effects are "fairly mild, and we have good measures to mitigate them....We have thousands of boats in the water at any given time, so there's already an enormous human impact on the ocean....We actually think we can be much more benign than that. So compared to the status quo, we think we can be very, very good in terms of environmental impact."

Brekken estimates that wave power costs approximately four to five times more than coal; however, he says this is where wind power was twenty or thirty years ago. Now, wind power is considered to be competitive with coal, especially if people consider the "full cost" of producing coal.

"Coal becomes a lot more expensive if you consider the ecological costs of acid rain," says Brekken. "How much is it worth for you to not have mercury in your water? Currently, a coal producer doesn't pay for those things, but they get socialized on the rest of society.... Once they start implementing in their plants the carbon sequestration capability—so they have to implement the various scrubbers and devices to capture the CO_2—it's going to get a lot more expensive."

Although he is quick to point out that he is not "down on coal," Brekken is most definitely up on renewable energy. And he looks forward to the day when wave power will be competitive with conventional technologies. He points to the lessons learned in the production of wind power, both from a technical and a political point of view: "We think we can learn from those lessons and we can accelerate that [learning] curve. So we can get somewhere closer to competitive energy costs faster than twenty or thirty years."

Like solar and wind power, wave power has low or no emissions. It is also relatively predictable and consistent, which solar and wind are not. Brekken believes the market for single-user applications, in isolated areas or on boats, may be as big or even bigger than the market for utility-size wave parks. Regardless of whether "they'd like to be able to drop some buoys off the side [of their boat] and generate some electricity" or whether they want to establish a larger wave park, Brekken sees more and more people looking at wave power as an important source of energy. He estimates it could provide as much as 10 to 20 percent of the "renewable energy mix" along the coastal areas.

Most utility companies use several types of energy—from coal to natural gas to nuclear to renewables—in their "power mix." Many list the percentage of each type of power—from coal to wind—that they purchase under "energy mix" or "power mix" on their websites. Some utilities offer "Green Power" programs where customers can choose to buy renewable energy power for anywhere from $5 to $10 extra per month (pennies per day). Virtually all utilities allow users to access their usage and payments in order to monitor and reduce their usage.

With an estimated 3 billion people (roughly half of the world's population) living in coastal regions, the potential of wave energy could very well turn out to be as big as the ocean itself, which is one reason Brekken says, "It's very important to continue to develop renewable energy." However, he warns, "It's just as important to continue to look at ways to improve efficiency and decrease our per capita energy usage."

4

Biofuel: Paradigm for the Future

If the twentieth century was the age of oil, then the twenty-first century will most certainly be the age of biofuels—at least if Larry Walker has a say in it. Walker is currently working with the United States Department of Transportation and the United States Department of Agriculture, the state of New York, and local business interests to design a "complete system" for the production and refining of cellulosic materials in the northeastern United States

"To a certain extent, [people are] kind of walking around with blinders," says Walker. "I think the energy situation could come back and bite us if we're not very careful. I think the global warming aspect could come back and bite us if we don't manage it carefully. And I think the water resources issue could come back and bite us.... And I think the time is *now* to make the investment in research and development and education to make this transition to sustainable biofuels, sustainable solar [and] wind, sustainable water usage."

A number of "peak oil" theorists believe that we have hit the tip of the proverbial petroleum iceberg and are now on the downward slope of oil availability. At the same time, our growing population is demanding more and more fuel to live and work and is gobbling oil, as one activist put it, "like two-legged SUVs." Under this bleak scenario, as an increasing demand overwhelms a steadily dwindling supply, oil will continue to become less and less available at any price. However, not everyone agrees that oil supplies have peaked. Opponents of the peak oil theory argue that there are plenty of reserves around the world—in places such as the Athabasca oil sands in Canada, and oil left in existing wells, oil which has heretofore been technologically unattainable.

In May 2008, oil prices spiked up to $135 per barrel, something which had once seemed impossible. Goldman Sachs analyst Arjun Murti wrote in a report at the time that "the possibility of $150–$200 per barrel seems increasingly likely over the next six to twenty-four months." In June 2008, the United States national average for unleaded gasoline broke the $4.00 mark for the first time in history. Rising oil prices are affecting more than just our mobility. They are affecting our ability to heat our homes and businesses, the cost of goods that we produce and consume, and even the price of our food. Dow Chemical announced it will raise its prices by up to 20 percent in an attempt "to mitigate the extraordinary rise in energy and related raw material costs." Other companies are poised to follow suit. An article published by *Bloomberg News* quoted analysts as saying, "With oil and commodity prices at record highs, these companies and others will be forced to pass on higher costs."

As fuel prices fluctuate, many consumers are actively looking for alternatives such as mass transit, alternative energy vehicles, and different types of fuel to decrease their oil and gas consumption and reduce their costs. Although a decrease in consumption may be beneficial in the short run, at some point, we will have to develop new technologies to supply the increase in demand. Biofuels are one source of alternative energy that has received a lot of attention lately.

Most people do not realize that, until the early 1900s, the average American household relied on a biofuel to heat their homes and to cook their food—they called it wood. In 1896, Henry Ford designed his first vehicle, the Quadricycle, to run on ethanol. And, by the year 1900, about a third of the cars in New York City were electric. The automobile was set up to be very eco-friendly, indeed—until huge oil reserves were found in Texas and California and cheap petroleum became available in the United States.

Biofuel: Paradigm for the Future

Petroleum is essentially ancient biomass—plant and animal material—that has decayed over millennia. Today, when we talk about biomass, we usually mean living plants such as corn or soybeans. However, biomass can also include recently dead or nonliving plant residue, animal matter, and biodegradable wastes. By the same token, when we talk about biofuels—fuels that are derived from biomass—we usually think of ethanol, but biofuels can also refer to other liquids such as biodiesel, gases such as biogas or synthesis gas, and solids such as wood or biomass pellets.

Biofuels can be used to power our vehicles as well as run our homes and businesses. And they are one of the main alternatives being considered to reduce America's dependence on oil, mainly because they do not require substantial infrastructural changes in the way we live and work. Consumers can still pull up to a regular gas pump to fill up; they will just be using E85 (a blend of 85 percent ethanol, 15 percent gasoline), or biodiesel, instead of gas or diesel.

"Ethanol is on the radar because it can fit into our transportation system today and in the foreseeable future, so that's driving a lot of the capability," says Walker. "From a science and technology standpoint, we can begin to engineer micro organisms, microbial communities, enzymes, systems to produce a broad spectrum of biofuels." The big question, according to Walker is, "How do we do it in a sustainable way? And I think that's the research challenge."

There are several processes by which people can produce biofuels, with new processes being introduced in each generation of development. Most so-called "first generation" biofuels have historically come from sugars, starches, oils, and fats such as those found in sugarcane, corn, soybeans, plant oils, and animal fats. In June 2007, the National Corn-to-Ethanol Research Center, a not-for-profit organization located in Edwardsville, Illinois, was quoted as saying that there were 110 ethanol plants in operation in the United States, with approximately eighty additional plants underway.

However, grain-based ethanol has come under a lot of fire recently for a variety of reasons, not least of which is that it simply may not be very efficient. According to an article in the October 2007 issue of *National Geographic*, it takes one gallon of gasoline to produce 1.3 gallons of corn-based ethanol. In addition, opponents claim any grain-based ethanol relies heavily on the same grains that help feed the world, resulting in higher prices for everyone from cattle ranchers to tortilla makers.

The EPA estimates Americans drive 231 miles per week (per vehicle) on average, which equals around 12,000 miles per year. In 2001, Dr. David Pimentel of Cornell University found that, "If all the automobiles in the United States were fueled with 100 percent ethanol, a total of about 97 percent of all United States land area would be needed to grow the corn feedstock. Corn would cover nearly the total land area of the United States." Dr. Pimentel was quoted as saying, "Abusing our precious croplands to grow corn for an energy-inefficient process that yields low-grade automobile fuel amounts to unsustainable, subsidized food burning."

Many researchers are now turning away from grain-based ethanol and exploring non-grain-based ethanol as a source of fuel. Second generation biofuels include cellulosic feedstock such as agricultural residue (e.g. orchard prunings or crop wastes), industrial and urban residue (e.g. wood pulp or construction project wood wastes), and perennial grasses (e.g. switchgrass). Much of the feedstock for cellulosic ethanol comes from waste by-products that would otherwise end up in landfills or be burned off in fields and create methane and carbon emissions.

Switchgrass has recently begun to generate much interest because its efficiency ratio could be much higher than that of corn—for example one gallon of gas is estimated to produce anywhere from two gallons of cellulosic ethanol to as many as thirty-six gallons of cellulosic ethanol (a 1:36 ratio). In addition, some studies indicate that switchgrass sequesters or stores carbon in its deep-root biomass, thus preventing CO_2 from dispersing into the atmosphere. Says Walker, "The notion that you can do some things to address climate change by recycling carbon, by getting higher CO_2 sequestration due to your agricultural activity—I think that's very exciting."

Walker envisions a "paradigm for the future," which includes a diverse portfolio of feedstock—regional grasses and other cellulosic material—including switchgrass, cold-tolerant sorghum, and woody biomass such as willow and hybrid poplar. He believes the greater the diversity of materials—each with its own growing and harvesting season—the easier it will be to regulate the flow of material to a bio-refinery over a 365-day period. While conventional grain farming can deplete the soil and use a great deal of water, switchgrass is said to prevent erosion and be water efficient. And, like trees, grasses do not need fertilizers (which are petroleum based and cause CO_2 emissions).

One of the logistical problems of developing a cellulosic bio-refinery is transportation. "We can ship corn across the country by rail, by barge, by ship because of its density," says Walker. However, he adds, "You're not going to be doing that with grasses; the densities will not allow us to do that. The transportation fees are going to eat you up." The obvious solution to this, according to Walker, is to focus on "regional production of the feedstocks and regional production of the fuel." An added bonus to this "regional distribution of energy" is that it provides an opportunity for rural communities to thrive economically. "One of the reasons that New York State is interested in this is, what does it mean for rural economic development?" says Walker. "And what does that mean in terms of generating jobs and economic opportunities?"

Walker does not think biomass is the only answer. "If we're able to get 30 percent of our transportation needs from biomass, we've done good," he says. He also believes it is crucial to "get higher levels of energy efficiencies out of our industrial systems [and] our domestic systems...if we can get the fleet mileage up to thirty miles per gallon, that would be fantastic in terms of saving energy, saving petroleum." In the end, Walker sees biomass as a part of an "energy matrix" that could include solar, wind, and perhaps even a "clean" coal technology.

Regardless of what part of the matrix biomass plays, Walker does not see it dropping off the radar like it did in the 1980s. "This whole concept of biofuels and bio-economies is not going to disappear like it did twenty years ago because there are some global drivers out there that will force us to continue down this path," says Walker. "Twenty-five years ago, China and India weren't the economic powerhouses that they are today. So there's more global competition for energy resources. Energy is essential for the quality of life that we live here in the West, and other parts of the world are going to want to have a similar life."

Like Dr. Walker, Dr. Jim Frederick first began looking into sustainable energies decades ago. As a chemical engineer working for General Motors in its polymer research department, he was asked to "take a look at how we can convert waste polymers into energy and fuels." From 2001 to 2003, Frederick held the Professor of Green Chemistry Chair at Chalmers University in Sweden, and his experiences there further strengthened his interest in renewable materials and energies. In 2003, he became director of the Institute of Paper Science and Technology (IPST) at Georgia Tech, which, in addition to its work with the paper industry, has also become a leader in biomass and biofuels research and development.

"We all come at our research from a 'this is cool science' point of view. I mean, there's no such thing as a researcher who doesn't do that. I think the question then becomes what, in addition to that, motivates you. And I come at it from a sustainability point of view as much as anything," says Frederick. This sense of a higher purpose has inspired him to look at the various ways cellulosic materials in trees—or in pulp waste generated from paper processing operations—might be used to reduce our dependence on fossil fuels.

As he explains it, there are two basic ways to convert biomass into biofuel: One is through biochemical conversion and the other is through thermochemical conversion. Sugarcane-to-ethanol is relatively straight forward to produce using a biochemical conversion process. Hot water is used to extract sugar from the cane, and the resulting liquid is heated and then fermented—much like beer—to produce ethanol. Corn-to-ethanol requires an additional step in which water and enzymes extract starch from the grain and then convert it into sugar. Cellulosic ethanol requires yet another pretreatment step to open up the cellulosic material and make it more accessible to the enzymes.

Frederick likens the process of extracting biofuel from cellulosic feedstocks to cooking spaghetti: "If you ever cooked spaghetti, and you dump the spaghetti in the water and don't stir it, what you get is the strand of spaghetti kind of congeals and gels together.... [Cellulose] comes as if it's those spaghetti strands all congealed together. And you can't get stuff in there. You can't get spaghetti sauce into [congealed spaghetti] and you can't get enzymes into [cellulose]."

The pretreatment stage needed to produce cellulosic ethanol is currently both costly and time-consuming, which is one of the barriers to cellulosic versus corn ethanol production. It is also one of the areas that Frederick and his colleagues are researching in order to develop faster and more cost-effective technologies for development.

Another area of research that Frederick is involved in has to do with the production of synthesis gas or "syngas" (a mixture of carbon monoxide and hydrogen) and other biofuels, which utilize a thermochemical rather than biochemical process. Thermochemical conversion involves the gasification of virtually any biomass into syngas, which, through catalytic conversion, can then be turned into a liquid fuel such as bio-methanol, bio-dimethyl ether (bio-DME), Fischer-Tropsch liquids, and mixed alcohols.

"The challenges of what we're looking at is it can be relatively difficult experimental work on the syngas production and conversion," says Frederick. "So, if we could get clever and find faster ways to do things, we'd be happier; we just haven't gotten there yet." Even so, he believes that these technologies could be ready for commercial use in "a year to years."

In addition to its work with government agencies, IPST works with corporations as well. Denny Hunter, vice president of technology at Weyerhaeuser, is chairman of the advisory board at IPST. Says Frederick, "We work very closely with our industry partners on, for example, the gasification of biomass and the clean up of syngas and also some ethanol related work. We've done quite a bit of work with these folks on projects where some of the companies are looking at possibly building an ethanol plant alongside their pulp mill so they can use some of the residual biomass to make ethanol and use waste-heat from the pulp mill to do the distillation and the evaporation that's required, because that's fairly energy intensive."

One of the ways IPST helps mills determine whether or not a project is viable is through a software package called BioRefinOpt which they developed at Georgia Tech. This program can be adapted to each individual facility's parameters to make predictions about possible yields and profits, among other things. IPST is also involved in the plant and process design and evaluation portion of a Department of Energy (DOE) consortium of universities which are looking at a pre-extraction process which would reduce the pulping process time for wood by a third and also produce ethanol as a by-product.

Also, Dr. Art Rigauskas, Georgia Tech's connection at the DOE's Oak Ridge National Laboratory, is working with scientists there to find ways to genetically manipulate different properties in tree embryos. Cellulosic biomass such as wood pulp contains cellulose, hemicellulose and lignin. Glucose or C_6 sugar (six-carbon sugar) is found in cellulose and can be fermented to ethanol using conventional yeasts. Non-glucose or C_5 sugar (five-carbon sugar) is found in hemicellulose and at this point in time cannot be fermented into ethanol with conventional yeasts on a commercially viable scale. Says Frederick, "They're basically starting from a genomics end and then [working toward] how to access the wood better with enzymes and how to get better enzymes and how to get the yeast that will convert the C_5 sugars from hemicellulose as opposed to the glucose that comes out of cellulose."

Unlike corn-to-ethanol bio-refineries, cellulosic bio-refineries are still in the planning and development stages and not yet up and running. A 2005 *BioCycle* article claimed that "only a few small demonstration biorefineries are producing ethanol from cellulosic feedstock," citing underfunding as one of the hurdles yet to be overcome. However, in 2006, President Bush announced his goal to make [cellulosic] ethanol "practical and competitive within six years" and the Biofuels Initiative was implemented to meet and to replace 30 percent of gasoline consumption with biofuels by 2030.

Says Frederick, "The corn ethanol is going well. The corn ethanol industry, if it started to use renewable energy for its process energy, would start looking a lot better [in terms of sustainability]. The wood ethanol one is [not] very far away. Nobody's got an operating plant yet, but in terms of technology I think we're pretty close on that. So I think six years is actually reasonable."

However, he adds, "I do think that in the long haul ethanol will be an interim fuel because, if you look at the miles-per-gallon per acre that you can get, ethanol is probably about a third of what we could get if we were making biodiesel or other diesel-like fuels. And that's a far more effective use of land and the resources that grow on it if we can go there. The trouble is those aren't six years away. Those are farther down the road."

In the meantime, says Frederick, "If we wait and sit back and say, 'Well, let's see what happens,' we may not be in a very good position to respond when something does." Frederick believes that it will take a combination of conservation and innovation to meet our upcoming energy demand, but he feels optimistic in the shifts that he sees in corporate attitudes towards renewable energies: "I think the fact that Chevron is looking at it as hard as they are— also British Petroleum has put a lot of money into it in this country—I think as an absolute minimum they're hedging their bets."

5

The Hydrogen Highway

Housed in the University of California, Irvine engineering laboratory facility, the National Fuel Cell Research Center (NFCRC) looks more like part of a modest manufacturing plant than a doorway to the future. But, as the oldest university fuel cell research center in the United States, NFCRC is at the forefront of what some have called an "emerging hydrogen economy."

The NFCRC is actually one of four component groups of the Advanced Power and Energy Program (APEP) at UC Irvine. Dr. Scott Samuelsen has been director of the program since 1970. Samuelsen looks like the quintessential college professor. He speaks precisely with a quiet voice that belies the dynamic nature of work that he has undertaken. Originally, he earned his Ph.D. in mechanical engineering at UC Berkeley and came to UC Irvine to develop more efficient and environmentally sensitive internal combustion systems. It soon became apparent to him, however, that he would need to explore more innovative ways to maximize efficiency and reduce waste.

"Going into the eighties, knowing that the combustion systems were going to have their limits...we began to look at an alternative technology," says Samuelsen, who originally looked into solar power before deciding that it, too, was limited due to the availability of adequate battery systems that could contain stored energy during periods of little or no sunlight. The NFCRC was born in the late 1990s, and scientists expected to have time to fully develop their ideas before they were thrust into the limelight.

"We did not expect the hydrogen era to take off as soon as it did, so that's kind of caught us off-guard," says Samuelsen. He credits the drastic acceleration to President Bush's "unexpected and unscheduled" announcement in favor of hydrogen-powered automobiles in his 2003 State of the Union Address and to California governor Arnold Schwarzenegger's campaign promise—and his subsequent follow-through—to develop a "hydrogen highway" in California.

"We are, I think, coming off of the hype that's [been] created from that," Samuelsen says with relief. "I think there's a Wall Street term that says that we are going through an 'adjustment.'"

With all the hype about hydrogen, it is easy to get facts confused with fiction. Even before nineteenth-century author Jules Verne wrote about hydrogen power in *The Mysterious Island*, scientists were experimenting with the gas. One hundred years before Verne's book, another Frenchman by the name of Jacques Charles launched the first hydrogen balloon flight. By the 1930s, hydrogen was used to inflate dirigibles for several trans-Atlantic voyages—until the Hindenburg went down in flames during an electrical storm. Although it was later determined that hydrogen had nothing to do with the Hindenburg disaster, people continued to view it as a dangerous, explosive gas. Hydrogen garnered little or no public support as a common source of fuel until the 1970s energy crisis got our attention.

In fact, most people don't know the difference between hydrogen and fuel cells. Hydrogen is the most abundant element in the universe. In its natural state, it is an odorless and colorless gas. When pressurized and cooled, it turns into a liquid. Hydrogen—whether in gas or liquid form—can be used to power internal combustion engines and has been used to power everything from rockets to fuel cells.

Fuel cells are devices that convert or "reform" fuel such as natural gas, methane, or hydrogen into electricity, with water and heat as by-products. Fuel cells act much like batteries in that they use anodes and cathodes to produce an electric charge. However, fuel cells don't store electricity like batteries do, so they need fuel to run continuously. Fuel cells may be either stationary (such as in a home or a building) or used for transportation or in portable devices (such as laptop computers and cell phones). Researchers are also exploring the use of microscopic fuel cells to power implantable medical devices.

Most of the talk nowadays has to do with hydrogen-powered fuel cell vehicles, which are being developed by a number of automobile manufacturers. But, as Samuelsen explains, "The future of fuel cells in the automobile is not yet certain."

Stationary applications are much more of a sure thing. They have been on the market for ten years and are just now starting to take off. Many of the stationary fuel cells today operate on natural or digester gas (a by-product of the anaerobic decomposition of waste), not hydrogen.

One example of a stationary application is the Sheraton Hotel in San Diego, California. Although powered by natural gas—a non-renewable energy source—the fuel cell system here still provides a boost in efficiency by using the "waste" heat and water vapor that is a by-product of fuel cell activity to pre-heat water for the hotel. Because no additional electricity is needed for this process, the amount of CO_2 emitted is reduced. According to Samuelsen, occupants can't tell any difference, but management can see a jump from 40 percent efficiency to 80 percent efficiency at no extra cost *and* with less environmental impact. In Samuelsen's words, "It's a win, win, win."

Another working example of fuel cell technology can be seen at the Sierra Nevada Brewing Company in Chico, California. In this case, the fuel cell system runs on digester gas, a by-product of the brewing process there. However, the best example of a "legitimate, renewable generation" source, according to Samuelsen, is the wastewater treatment plant in Santa Barbara, California, which powers its fuel cell system with digester gas produced during waste processing. In this case, the digester gas provides the fuel to power the cell, which then produces electricity, and the energy is generated with almost zero emissions. The heat from the fuel cell is captured and used to support the digestor reactor.

Global Warming I$ Good for Business

Unless you are specifically looking for it, it would be difficult to tell whether the building you are in is powered by a gas turbine or a fuel cell system. Stationary fuel cells, which are a form of distributed (versus central grid-connected) energy, are often housed in standard utility cabinets, like any other on-site power generation system. But, unlike traditional gas turbine systems, fuel cell systems can use their by-products—namely waste heat and water vapor—to power equipment. In addition to preheating water (like the Sheraton Hotel system), the waste heat and water vapor expended by the fuel cell can also be used to power an absorption chiller and thus provide air conditioning for a building facility. The NFCRC installed this type of system in January 2008.

Although he is a proponent of fuel cell systems, Dr. Samuelsen doesn't believe that we have to use *either* fuel cells *or* gas turbines, which both use internal combustion technology. He envisions utilizing the best of both worlds by using both types of systems to power a variety of machines, including aircraft, ships, locomotives, or even long-distance transport trucks. In the case of a ship or an airplane, for instance, the internal combustion gas turbine engine could be used for propulsion, while a fuel cell could be used to provide the "hoteling" power to operate the lighting, the air-conditioning, and the controls.

Stationary hybrid fuel cell systems are one of the main areas of research at UC Irvine, where fuel cells are integrated with a gas turbine engine. The development of the hybrid is a "huge technology," which produces "high efficiency and almost zero emission of any criteria pollutant," according to Samuelsen.

A photovoltaic/fuel cell hybrid system is even more environmentally friendly since it is both renewable and regenerative. The way it works, in theory, is this: At night, when the sun is down and you are home, the fuel cell runs in the "forward" direction, producing electricity, heat, and water. During the day, when the sun is up, the solar panels provide electricity, which runs the fuel cell in "reverse" order, taking the water and turning it into hydrogen and oxygen through a process called electrolysis. The hydrogen is then collected, stored, and used to power the cell in a "forward" direction again in the evening, when the sun goes down. This regenerative, renewable, and non-fossil fuel system could potentially be used in both homes and businesses and is essentially a zero-emission 24/7 system.

Samuelsen estimates that it could take more than five to ten years to develop such a technology; however, he sees great promise for this type of "reversible" fuel cell in Honda's partnership with Plug Power in New York. At the end of 2007, Honda announced that it had begun using the Home Energy Station IV, a fourth generation unit "designed to provide fuel for a hydrogen-powered fuel cell vehicle [such as the FCX Clarity], as well as supply heat and electricity for a home." Samuelsen hopes that Honda's vision and visibility can spur further research and development in this area.

One of the main barriers to the use of hydrogen is a lack of infrastructure. Although natural gas lines are commonly run to businesses and residences, hydrogen gas lines are not. In terms of transportation, fueling stations for vehicles are few and far between. One of Samuelsen's jobs has been to work with the Hydrogen Highway Network staff to help determine how and where to deploy hydrogen stations throughout California. Originally, they considered placing stations every twenty miles between Los Angeles and San Francisco; however, they eventually decided to concentrate on urban areas and then branch out to more rural areas from there.

Another area of concern for Dr. Samuelsen is producing hydrogen in the most "environmentally responsible way." Hydrogen gas has a very low density compared to other gases, and energy is required to condense it into a liquid. Critics claim that other technologies such as the electric vehicle (EV) or even the plug-in hybrid vehicle (PHEV) are more efficient and environmentally sound. However, Samuelsen thinks that these critics are missing the point. He believes alternative fuel vehicles "enable the hydrogen future because the hydrogen fuel cell vehicle that we have…is going to benefit [from] hybrid battery-electric capability…just as the Prius does today." He sees hybrid hydrogen systems as the best-case answer to our transportation needs.

UC Irvine is currently working in collaboration with several private and public partners, including Toyota, General Motors, the California Air Resources Board, and the South Coast Air Quality Management District. One of the goals is to research and develop the use of a PV/fuel cell system that can be used to power plug-in electric vehicles, which can in turn be used as part of an overall transportation system designed to reduce traffic and emissions.

The Zero Emission Vehicle-Network Enabled Transport (ZEV-NET) system that has been developed at UC Irvine works like this: Let's say you take the train into Irvine, California (or New York or Chicago or any other city). You arrive at the station in the morning, use your "smart card" to pick up one in a fleet of ZEV-NET vehicles and drive to work. While there, either you or any of your coworkers use the vehicle to drive to-and-from your respective business destinations. At the end of the day when you get off work, you pick up your ZEV-NET vehicle and drive back to the train station, where you plug it in. Then you take the train home for the day.

According to Dr. Samuelsen, this program accomplishes two things: It reduces CO_2 and other emissions from vehicles and it combines the best of public and private transportation—it is user-friendly, convenient, and helps reduce traffic in an area known for its freeway gridlock. And, while ZEV-NET does not yet utilize PV solar technology to power the vehicle recharging, it is actually in operation in Irvine, California.

Hydrogen gas, which is measured in kilograms, currently costs $12 to $13 per kilogram. With subsidies, the end-user at UC Irvine's hydrogen fueling center pays $5 per kilogram, which, according to Samuelsen, is roughly equivalent to paying $2.50 per gallon of gasoline because of the increased fuel efficiency. Samuelsen believes the cost of hydrogen will decrease as technology and infrastructure are developed. In addition, as the price of gasoline continues to rise, he believes hydrogen fuel will become a viable alternative for internal combustion engines.

The cost of fuel cells will also go down, according to Samuelsen. Current proton exchange membrane fuel cells—a type of fuel cell that can be used in both transportation and stationary fuel cell applications—are expensive. However, he explains there have been several recent breakthroughs in technology. "There's enough going on and enough results coming out of [it] that I don't see that as a problem," says Samuelsen. "I do see the naysayers and others raising that [objection] right away because this is all new technology and it's easy to do. It's got to cost a lot, how do you mass-produce that and so on and so forth. But now there is a pathway that's very clear to me, and we are going to work through it to get to a reasonable price."

In the meantime, Samuelsen says, budding renewable technologies from biodiesel to electric vehicles—while they may not be a cure-all in and of themselves—can still have a beneficial effect on the environment and will ultimately help raise a "renewable consciousness" among the public. Without that shift in consciousness, Samuelsen says, nothing will change:

"The public, in the end, has to be behind this in one way or another...when the public starts to get around it, then the businesses start to feel more comfortable and certain [government] policies enable the businesses to begin to work in this space more comfortably." He adds, "The opportunities for growth are unlimited, but they come with the recognition that the earth can no longer accept and handle the type of products that we have had in the past and the way we have done business in the past with respect to the generation of electricity and meeting our transportation needs."

6

Eco-Capitalists

Dr. Tony Michaels is a "pracademic"—a practical academic with the mind of a scientist and the heart of an entrepreneur (or maybe it is the other way around). He sits on the edge of his seat, eyes sparkling, with a barely controlled sense of excitement that is infectious. "Much of society is driven by fear-based responsiveness," he says. "And we [here at USC] just think that it makes no sense; so we have an opportunity-solution kind of focus."

Michaels has been at the University of Southern California since 1996, and his office is a stone's throw away from one of the oldest and (as of 2007) top-ranked graduate entrepreneur programs in the United States. In a university that has made a point of nurturing innovation to implementation, he has put together an interdisciplinary team whose focus is on not just studying but actually carrying out solutions to environmental problems. He believes "environmental entrepreneurship" is the answer to some of today's most pressing environmental and energy-related problems.

An oceanographer by training, Michaels wrote his dissertation on "Global Warming and the Role of Oceans in Taking up Greenhouse Gases" at the University of California, Santa Cruz. However, unlike most academics, he also speaks passionately about the positive effects that can occur when scientists and entrepreneurs get together to bring cutting-edge research to the marketplace.

"It's a mind shift," says Michaels. "The most effective thing we can do is create some interesting nugget of solution and get it in the hands of somebody that's going to use an entrepreneurial approach to get rich off of it. And, when you do that, the more success they have, the more that solution spreads around the world."

Global Warming I$ Good for Business

According to Michaels, one of the barriers to getting good ideas off the drawing board and into society is what he calls the "valley of death"—that period of time after the research funding has run out but before the project is commercially viable. To deal with this gap in funding, Michaels says that he and his colleagues are endeavoring to start a venture fund which will enable a wide range of individuals and organizations to invest in some of the high-risk, nascent technologies coming out of universities across the country.

According to Michaels, the fund will start with four diverse technologies, all of which are in the advanced research stage. The first is a type of microbial fuel cell, which uses "biofilms" that consist of colonies of bacteria and other microbes to treat sewage without producing sludge and then produces electricity from the by-products. "Conceptually, it's done," says Michaels. "And it scales from [the size of a coffee mug] to municipal size."

He hopes the value of sewage treatment alone will make this a worthwhile venture, both in terms of profit and in terms of societal value. "The fact that you can treat sewage with no sludge is a win in its own right," says Michaels. "[There are] two billion people without adequate sanitation and drinking water because there's waste in it. You don't have to change their practice; you make the waste valuable so they don't dump it in the stream, they strip the electrons out of it."

Another technology that is being considered for the fund is similar in practice to the microbial fuel cell in that it utilizes microbes to strip certain elements out of water. In this case, though, the microbes are genetically manipulated to strip arsenic and chromium out of drinking water at a very low cost (according to Michaels, scientists have also developed a variant that will strip uranium out of water). Again, Michaels sees some far-reaching societal benefits from this technology, especially in terms of cleaning industrial waste out of drinking water: "[There] should be a market of some sort for this kind of thing, and it could be quite large because you're talking about extraordinary numbers of people that have compromised health because of it."

Michaels hopes the microbial fuel cell and the toxic metals technology will be available in three to five years. However, he believes the proposed fund's third technology—carbon sequestration—may be available before that.

Carbon sequestration is the process of removing carbon dioxide from the atmosphere either through artificial capture and storage technologies or through natural sequestration. The Department of Energy calls it "one of the most promising ways for reducing the buildup of greenhouse gases in the atmosphere." Oceans, plants, and other organisms that use photosynthesis to remove carbon from the atmosphere and then incorporate the carbon into biomass are all examples of natural carbon sinks, or reservoirs.

Although different methods of ocean sequestration are currently being studied around the country, the USC group is focusing on "nitrogen fixation," a process that stimulates plants with iron particles and creates new nutrients from nitrogen gas. "If you stimulate this kind of biology, it stores carbon, and it stores carbon for hundreds of years," says Michaels. "But I can't point at ten studies that say, if you dump iron on the surface, this thing grows." What is needed now is more research and development to measure the reliability of this methodology and the environmental impacts that it might have.

The fourth technology to be researched and possibly developed is robotic aquaculture. Farmed seafood, according to Michaels, "is the only expanding source of high-quality protein on this planet." Since most of the coastal areas around the world are already taken, Michaels says, his group has proposed using robotics to automatically tend deep sea fish farms.

"We have a dual approach here," says Michaels. "One company is looking at the robotics, the automated tending of fish farms, and the other is looking at the genetics of things that are farmed." Although genetic modification remains a controversial subject, Michaels claims that his approach is much like the selective breeding that human beings have been doing for generations with crops and livestock.

Perhaps one of the most innovative things that Michaels and his team are working on is where to get funding for research. They have taken the idea of private endowments one step further to develop what Michaels calls "venture philanthropy."

In this type of arrangement, those organizations and individuals who might otherwise have gifted money to a university for research and development can now invest those same dollars in a venture fund. Says Michaels, "Instead of making a gift, make an investment...the worst that can happen is good science gets done and the write-off for a business loss has the same implications as a donation." However, if, as he hopes, the fund does well, Michaels expects everyone to make a very nice return on their investment.

Another benefit of utilizing a venture fund for research is that professional hedge funds, which "can't just do philanthropy," can invest a portion of their portfolio in the fund as well. Although Michaels admits there are "extraordinarily high risks," he also believes there is an "extraordinarily high potential reward."

Whether or not Michaels will succeed remains to be seen. He admits that being both research scientist and entrepreneur can lead to conflict: "We want to be the most faculty-friendly, university-friendly game in town but at the same time manage the conflict of interest, conflict of commitment—all of the fundamental issues of why a university is a not-for-profit entity out there." According to Michaels, one way to do that is "being willing to walk away from some of [the viable business projects] because it's more important not to wreck the academic connection than...just to get one product out the door."

For Michaels, the biggest barrier to innovation is bureaucracy. "Most academics use the science we produce to guide a regulatory process," he says. "And, yeah sure, it needs to be done. You know, there're real rules for real reasons...but it's such a limited approach. It's always about constraining the actions of others." Instead, Michaels suggests, "Go out there and create solutions and use the positive force [of the free market] to spread them around."

The emergence of green technology in the information age has spawned a new term: clean technology. Often used interchangably with the the term "green tech," clean technology or "cleantech" includes biofuels, geothermal, hydroelectric, tidal/wave, fuel cells, solar, wind, and other renewables. However, cleantech is more than green, according to the Cleantech Network: "Cleantech is new technology and related business models offering competitive returns for investors and customers while providing solutions to global challenges." The Clean Energy Patent Growth Index, published quarterly by a New York law firm that handles tech patents, estimates approximately 900 cleantech patent applications were granted in 2007—up from approximately 750 in 2002.

The University of California, Davis has a unique approach to helping cleantech innovations get to market, which is based on the concept of social networking. Every year in July, the UC Davis Center for Entrepreneurship offers a five day "immersive" Green Technology Entrepreneurial Academy (Green TEA) to science and engineering doctoral students, post-docs, and faculty. Green TEA is designed specifically to help technology-oriented individuals network with business experts and venture capitalists. According to Nicole Starsinic, assistant director for the Center for Entrepreneurship, the goal is not so much to match individual inventors with potential investors (though that may happen) but to "bring researchers in science and engineering, working on sustainable and green technologies, together with a network of investors and industry executives to help them gain an understanding of the network needed to bring ideas into the marketplace."

The academy, which was initially developed for UC Davis faculty and students, is now open to green tech scientists and researchers from all over the country. The participants work in groups of three or four to jointly develop the business potential of each of their research projects. At the end of the week, they pick one project to pitch to venture capitalists, angel investors, and "serial entrepreneurs," who give them feedback and mentor them in their format and presentation. Project proposals are reviewed for their market interest, competition, technological strengths and weaknesses, and how likely the technology is to succeed.

According to the Green TEA brochure, over the course of five days, participants learn to "navigate the path from research to market, manage the dynamics of entrepreneurship, evaluate technology/market opportunities, pursue patent and licensing strategies, write and communicate business plans, build and manage interdisciplinary teams, and find and tap funding opportunities." Thus far, response to the Green TEA program has been very positive, as more and more "techies" become interested in getting their ideas to market.

While some participants leave the academy to start their own cleantech companies, others decide to develop their technologies from within or license them to larger corporations. Still others utilize what they have learned to "refine their research." Regardless of which path they choose, says Starsinic, participants should come away from the academy with a better understanding of what they will need to do to transfer their technologies to market and also—perhaps even more important—they will have a list of contacts who can advise and mentor them through the process.

"The technologies range broadly," says Starsinic. "There is research into using algae as biofuels, there's work in fuel cells, solar cells, and energy efficiency." One example is a technology to reduce energy consumption through home energy usage monitoring. "This is a growing niche that the utilities are interested in that merges green technology with information technology," says Starsinic.

Dr. Andrew Hargadon, director of the Center for Entrepreneurship, is the driving force behind Green TEA and is also the founding director of the Energy Efficiency Center at UC Davis. For Hargadon, the issue goes beyond developing new technologies. In his words, "We may need revolutionary new technologies to save us from our old ones, but we also need revolutionary new ways of changing."

Usually, new ways of changing require looking at old ways of doing things and then coming up with something that addresses a fundamental flaw in the system or fills a need that no one ever thought of filling before. Often, the concept is so simple that, when others hear of it, they wonder why they didn't think of it themselves.

In 2006, Dr. Ruihong Zhang, a professor of biological and agricultural engineering at UC Davis, announced a new technology called an "anaerobic phased solids digester" which was licensed by the university for commercial use to Onsite Power Systems, Inc. The Biogas Energy Project, as it has been called, turns table scraps from Bay Area restaurants into clean energy which could potentially provide electricity to power ten average California homes for one day.

In anaerobic digestion, organic materials such as food waste are mixed with bacteria and put into an airtight container. Over a period of time, the waste is broken down into sugars which are then converted into organic acids and finally produce a biogas such as methane that may be used for heating or generating electricity. Although anaerobic digesters have been used at municipal wastewater treatment plants and livestock farms for some time, Dr. Zhang noted that they did not process the variety of wastes (i.e. both solids and liquids) as quickly as the Biogas Energy Project, nor did they produce both hydrogen and methane gas.

Onsite Power Systems CEO Dave Konwinski reportedly predicted, "This technology will make a substantial dent in both our landfill needs and our use of petroleum and coal for fuels and electricity. It also will reduce our greenhouse gas emissions."

The idea of generating a product or service from waste is, of course, not unique to UC Davis. Rutgers' EcoComplex, a member incubator of the Clean Energy Alliance, has also facilitated the development of ideas along those lines. One example is a simple, down-to-earth product made by TerraCycle, Inc.—worm poop lawn and garden fertilizer.

TerraCycle cofounders Tom Szaky and Jon Beyer wanted to develop a company that was "financially successful while being ecologically and socially responsible." The two Princeton students discovered that there was money to be made in "eliminating the idea of waste."

TerraCycle's "all-natural, all-organic, 'goof-proof' liquid plant food is made from waste (worm poop) and packaged in waste (reused soda bottles)" and is the first in a line of reused waste products that is rapidly expanding to include wine barrel composters and rain barrels (made from used wine barrels), planters and "art pots" (made from compressed computer parts), and even handbags and totes (made from those ubiquitous juice pouches found in schools across the country).

According to company spokesperson Jennifer Wilkie, "Recycling is good, but reusing is better because it takes less energy, etc., to reuse something than to recycle....The goal is to eliminate waste by turning it into something useful. Every year we're adding new products because there's just so much garbage."

In a February 2008 TreeHugger.com article, Szaky claimed, "This idea, called 'sponsored waste,' benefits any large company with a non-recyclable packaging and helps TerraCycle provide consumers with affordably priced, eco-friendly products." In essence, TerraCycle is taking waste off the hands of producers such as Stoneyfield Yogurt, Cliff Bar, and Capri Sun and then reusing that waste to produce its own products.

TerraCycle began this process by enlisting the help of four schools in collecting bottles for its fertilizer. Now, its "brigade program" consists of around 4,000 schools and community groups who help collect a variety of waste items for reuse. For example, Nabisco, maker of Oreos and Chips Ahoy!, is part of TerraCycle's "cookie wrapper brigade." TerraCycle sends out collection bags to individuals or groups who are interested in joining the brigade to collect the cookie wrappers (there must be hundreds at any given school). Nabisco donates $0.02 per cookie wrapper to the charity or school of the collector's choice. TerraCycle then cleans the items and uses them to make its own high-quality, cost-effective products. It is a win for everybody and a clever business model. Said Szaky, "TerraCycle has a unique opportunity to help larger companies to reduce their waste streams while procuring zero-cost materials to make eco-friendly products."

You would think that products made of reused garbage could only be sold at fringe outlets frequented by hard-core environmentalists, but TerraCycle products are sold to mainstream markets at Walgreens, Target, and Home Depot. Apparently, customers actually like the products in and of themselves, not just because they are environmentally friendly. In May 2008, TerraCycle reportedly teamed up with OfficeMax to sell eco-friendly office products such as innovative "green" binders, pencil cases, trashcans, and cleaners at OfficeMax stores. As with all TerraCycle items, its packaging and products are made entirely from waste.

All of this might seem like heady stuff for two former college kids, but they seem to be right on track to achieving their objectives: They have produced a product that is ecologically sensitive and socially responsible and financially successful; in other words, they have met their "triple bottom line."

In his "Eco-Capitalist Guidebook," Szaky posed the question that is, perhaps, at the core of cleantech development today: "What if we could achieve a triple bottom line report card where each bottom line drove the next and everybody won?...The implications would be tremendous."

Part Two
The Game Changers

7

The Environmental Power Players

In the 1960s and 1970s, it would have been inconceivable for corporate executives to join forces with environmentalists. However, today's multinational corporations are joining forces with nongovernment and nonprofit organizations in what looks like a twenty-first century version of a "love in." Their goal: to develop ideas and strategies for dealing with climate change and energy shortages.

These companies understand that they are going to have to meet global standards for sustainability if they want to compete in a global economy. Like so many others throughout the world, they recognize the time has come for the United States to take the lead and move forward with a clear-cut plan of action. Many recall the race to the moon during the 1950s and 1960s and the amazing technologies that were developed by the free market, with clear guidance and incentives on the part of the U.S. government, and they are calling for a similar approach to climate change.

In January 2007, the United States Climate Action Partnership (USCAP) issued a statement urging U.S. lawmakers to enact a policy framework that would, among other things, "reduce global atmospheric GHG [greenhouse gas] concentrations to a level that minimizes large-scale adverse impacts to humans and the natural environment." The group recommended establishing emission reduction targets as well as a national program to accelerate technology research, development, and deployment. USCAP also recommended encouraging other countries, including those in the developing world, to implement GHG emission reduction strategies.

Not surprisingly, founding members of this group consisted of such groups as the Environmental Defense Fund, the Natural Resources Defense Council, and the World Resources Institute. However, somewhat more surprising, founding members also included several unlikely "environmentalists" such as Caterpillar, Duke Energy, DuPont, and other major manufacturers and industrialists, and the list is still growing.

After a year of discussion, USCAP has come up with its own recommendations "to guide the formulation of a regulated economy-wide, market-driven approach to climate protection." One of the key recommendations is a cap-and-trade system that will, in the words of John W. Rowe, chairman, president, and CEO of Exelon Corporation, "provide regulatory certainty and create economic opportunity." Said Rowe, "If we do it well, carbon regulation can serve as an incentive to the next great global industry—low carbon energy production and efficiency. If we do it poorly, carbon regulation can impose an impossible burden on our economy and seriously undermine our global competitiveness."

In USCAP's 2007 report, "A Call for Action," the group presented two possible suggestions for implementing a cap-and-trade program. The first—"an upstream program"—would require fossil fuel producers to buy allowances for emissions released when the fuel is combusted. The cost of the allowances would be added to the cost of fuel, which would reflect more of a "full cost" than is currently in place.

The second program—"a hybrid program"—would include a "downstream" cap on greenhouse gas emissions from large stationary sources and an "upstream" cap on the carbon content of fossil fuels used by other remaining sources. Both approaches would set specific, federally mandated limits on overall annual greenhouse gas emissions.

In June 2007, Mark MacLeod, director of special projects for the Environmental Defense Fund (the U.S.-based, nonprofit environmental advocacy group), explained that the cornerstone of any cap-and-trade policy is the cap—"an absolute, nationwide limit on the pollution that causes global warming." According to MacLeod, "The 'trade' part is a market that creates powerful incentives for companies to reduce pollution, and provides flexibility in how companies can meet the limits."

Simply put, the government sets a cap on the amount of annual CO_2 emissions and then creates allowances equal to those emissions. Companies can choose to reduce emissions enough that they have more allowances than they need, in which case they may sell their allowances at a fair market price. Alternatively, companies may decide that it is more cost-effective for them to buy allowances than it is to reduce their emissions, in which case they may purchase allowances from other companies or from a middleman.

In 2003, the Chicago Climate Exchange was launched as "the world's first and North America's only active voluntary, legally binding integrated trading system to reduce emissions of all six major greenhouse gases, with offset projects worldwide." Current members include such big name corporations as Ford Motor Company, Dow Corning, and Monsanto, as well as various cities, states, and universities.

According to MacLeod, "The key advantage to a cap-and-trade system is that the more a company reduces its emissions, the more money it can either make or save." However, he added, "One limitation of cap-and-trade is that it should only be used for certain kinds of pollution." For example, CO_2, which is widely dispersed throughout the upper atmosphere, is a "good fit" for cap-and-trade, while mercury emissions, which are deposited within a limited local area, are not.

Not everyone prefers a cap-and-trade system, however, and some have questioned the motives of USCAP. In 2007, Charles Komanoff, cofounder of the Carbon Tax Center, wrote, "USCAP is looking less and less like a CO_2 control lobby and more like a corporate club seeking to cash in on the rising clamor against free carbon spewing." Komanoff criticized the cap-and-trade system as being "devilishly complex" and recommended instead a straightforward tax on the carbon content of fossil fuels that is "simple, transparent, and equitable."

According to Komanoff, "A carbon tax actually provides more precise price signals, provides them sooner, and provides them in a more understandable and transparent fashion." Furthermore, he wrote, "Carbon taxes will require no new administrative structures, can be implemented now, and can be adjusted as necessary." In addition, he claimed, "Carbon taxes are relatively immune to manipulation [by special interests]."

The Carbon Tax Center advocates "phasing in" carbon taxes over time in order to give businesses and households time to adapt and to develop new technologies and methods of operation that are less carbon intensive. In addition, the group believes that carbon taxes should be "revenue-neutral," meaning "the vast majority" of carbon tax revenues will be returned to the American people and not retained by the government. One way to ensure this is through a rebate program similar to the one in Alaska, where citizens receive an annual dividend. The other is known as "tax shifting," where any increase in carbon taxes will trigger a reduction in other taxes (such as sales or payroll taxes). According to the Carbon Tax Center, Sweden, Finland, Norway, and the Netherlands have all introduced a carbon tax. "Ironically," according to the Carbon Tax Center, "The Kyoto Accords halted further enactment by individual countries, but the recent upsurge of climate concern has brought a resurgence of interest in taxing carbon."

New York City Mayor Michael Bloomberg has spoken in support of a national carbon tax on several occasions. In his November 2007 testimony before the House Select Committee on Energy Independence and Global Warming, he said, "As long as greenhouse gas pollution is free, it will be abundant. If we want to reduce it, there has to be a cost for producing it—which means putting a price on carbon *indirectly*, through a cap-and-trade system, or *directly*, through a charge on all carbon use. The primary flaw of cap-and-trade is economic—price uncertainty, which could have harmful economic effects; while the primary flaw of a pollution fee is political—because proposing new fees is unpopular."

According to Bloomberg, the costs of a carbon tax or cap-and-trade are the same (and may actually be more with cap-and-trade). Ultimately, we will have to make a choice. "Whichever route we choose," he said. "We can't be afraid to act. Global warming is testing America's leadership on the international stage, and it is testing our resolve here at home."

This need to act is being felt by even the most hard-nosed business people, and many are making a voluntary choice to reduce greenhouse gas emissions. Some are increasing their energy efficiencies and reducing their waste, while others are implementing new technologies such as on-site solar or wind generation. Whether their motives are altruistic or capitalistic, whether they are acting in order to preserve the planet for their children or simply to make a buck, more and more companies seem to be putting their money where their mouth is when it comes to making the changes necessary to be both economically and environmentally sustainable in the twenty-first century.

Multinational corporations often exhibit symbiotic dynamics similar to those found in natural ecosystems, with technological infrastructures and social networks that are miles apart and yet are economically interdependent. These corporate ecosystems may consist of multiple business units, with headquarters on one continent, manufacturing facilities on yet another, and employees sprinkled around the globe. As the world shifts its standards for greenhouse gas emissions, these companies must shift their actions as well, not only to protect the environment but also to protect their market share and grow profits.

Many of these organizations are focusing on the same "stabilization wedges" that Princeton's Robert Socolow and Stephen Pacala defined in 2004, with an emphasis on building more efficient vehicles, more efficient buildings, more efficient power plants, and, of course, developing alternative and renewable energies. At the core of these innovations are so-called "intrapreneurs"—those individuals whose job, within an organization, is to think outside of the proverbial box and look at new ways of doing old tasks.

While corporate America gears up to deal with climate change, small businesses around the country are busy filling market niches that giant organizations either cannot or will not address. The Center for Small Business and the Environment (CSBE) says that small businesses constitute one half of our country's economy, creating two-thirds of all new innovations and virtually all new jobs, making them "the best source of nonbureaucratic solutions to environmental problems."

The CSBE is founded on the belief that "profitable, efficient, and innovative small businesses can lead the way to a new economy that protects and restores the environment while it produces abundant growth and employment." The idea of entrepreneurs laying the foundations of tomorrow's growth is nothing new. Entrepreneurs and inventors such as Thomas Edison and the Wright brothers have certainly laid the foundation for growth in this country. What many people do not realize are the many failures these inventors encountered along the way. Edison, for example, failed thousands of times before designing a functional incandescent lightbulb. And the Wright brothers (both of whom were high school dropouts) worked for years before their first successful flight, which lasted only twelve seconds and traveled a mere 120 feet.

It is easy to become impatient with new attempts at change because most of us do not like change in the first place. We may see the need to do something different, but we do not really want to go there. Entrepreneurs not only see the need, they embrace the changes necessary to find new, often disruptive, technologies which make the current market obsolete.

Today's cleantech inventors and entrepreneurs have the same spirit as their predecessors. They see in climate change not a threat but an opportunity—a need that they can fill. However, unlike their predecessors, these men and woman are often highly educated and sophisticated in their approach. They come from different countries, different walks of life, different cultural identities, but they have a similar goal—to change not only their position in the game but to change the very game itself. If they are successful, they will not only make a tremendous profit, but they will alter the way we live and do business in the next century and beyond.

With the green economy projected to grow, by some estimates, to more than $4 trillion, large and small companies alike are finding tremendous opportunities not only to make a profit but also to transform the very nature of their businesses. However, in the melee of environmental capitalism, it is important for companies and consumers alike not to get caught up in the hype of advertising products that are not as green as they may seem. "Greenwashing" is a type of advertising spin which organizations may use to convince consumers that their products or services are more environmentally friendly than they actually are. Some companies do this on purpose, but a great many get caught in their own hype.

TerraChoice Environmental Marketing lists "Six Sins of Greenwashing": the sin of the hidden trade-off, the sin of no proof, the sin of vagueness, the sin of irrelevance, the sin of fibbing, and the sin of the lesser of two evils. According to TerraChoice, hidden trade-offs are when companies tout a product as being sustainable but fail to mention the environmental impacts of production and transportation. For example, saying that a given paper product is environmentally friendly because it comes from a certified sustainable forest, while failing to mention the environmentally unfriendly way in which it is produced or transported, is a hidden trade-off. "Emphasizing one environmental issue isn't a problem," says TerraChoice. "The problem arises when hiding a trade-off between environmental issues."

By the same token, companies often make claims but offer no evidence or certification by a third party to verify those claims, or they make claims that are so vague that they are impossible to verify and are therefore meaningless. Sometimes companies make claims that have nothing to do with the product they are selling (who cares if your liquid soap is CFC-free?) and sometimes they outright "fib" about the environmental benefits of their products, usually by claiming they have third-party certifications when they do not. Lastly, some companies try to make consumers believe that a product is "green," when it is actually of questionable benefit (e.g., organic tobacco for cigarettes).

In its recommendation for marketers, TerraChoice states, "Green marketing is a vast commercial opportunity, and should be. When it works—when it is scientifically sound and commercially successful—it is an important accelerator toward environmental sustainability....Green marketers and consumers are learning about the pitfalls of greenwashing together. This is a shared problem and opportunity."

8

The Nuclear Reactor in Our Backyard

When we think of saving the planet, most of us do not think of nuclear energy. For many, nuclear power is synonymous with Three Mile Island and Chernobyl, with visions of men in white hazmat suits going into defunct reactors or of frightened children clutching their parents' hands while tearfully leaving the only home they have ever known. However, proponents today are touting nuclear energy as the safest, most reliable, lowest carbon generating source of power available. They say if our goal is to reduce greenhouse gas emissions *and* meet our country's growing need for energy, then nuclear power is the best way to go.

The International Atomic Energy Agency published a bulletin in 2000 that compared the total life-cycle greenhouse gas emissions of nuclear energy to other forms of energy and found the life-cycle emissions for nuclear energy were less than wind, PV solar, or hydro power and significantly less than coal, oil, or natural gas.

"There is no practical way to meet the challenge of designing long-term measures to reduce greenhouse gases without nuclear power being a part of the future," said Edison International chairman, president, and CEO John E. Bryson in 2007. Bryson echoed the sentiments of many industrial and political leaders and even a few environmentalists in extolling the virtues of this technology.

The United States has over 100 nuclear reactors, all of which are a form of light-water reactors. The United States Nuclear Regulatory Commission says the two basic types of reactors currently in operation in the United States are boiling water reactors and pressurized water reactors. Both types of reactors rely on steam to run a turbine, which in turn runs a generator to produce electricity, and both are cooled by water.

Visitors to Southern California have probably never heard of the San Onofre Nuclear Generating Stations (SONGS), located just south of "the OC" in San Diego County. SONGS utilizes two pressurized water reactors and is capable of providing enough power for 2.2 million households. Jointly owned by Edison International subsidiary Southern California Edison (SCE) and several other utilities, SONGS reactors have provided steady, reliable power for forty years, thus avoiding the production of approximately 100,000 metric tons of smog-producing air pollutants and 180 million metric tons of greenhouse gas emissions, according to SCE.

Many people do not realize that nuclear facilities, such as the one at San Onofre, have been a constant source of electricity in the United States for decades (SONGS has been in operation since 1968). One of the biggest draws for this type of power is that it supports grid reliability. According to SCE, "Electricity providers face the same challenge as water providers, maintaining both an adequate supply and the needed 'pressure' to deliver that supply....The San Onofre facility provides essential voltage support to Orange and San Diego counties."

Anyone who has experienced a brownout or a blackout can appreciate the need for consistent, reliable energy. Nuclear power plants can provide baseload generation that is available 24/7, regardless of variables such as weather (a limiting factor for solar or wind). In addition, nuclear power plants do not emit greenhouse gases when they generate electricity. Currently, more and more people are considering nuclear power as a viable alternative to fossil fuel power plants.

Duke Energy describes the generation of electricity in a nuclear power station as being similar to a coal-fired steam station, except that coal-fired power plants burn coal to produce heat, while nuclear power plants split uranium atoms. In nuclear fission, the uranium atoms are split (i.e. fission) by neutrons. Once split, the neutrons from these atoms collide with and split other atoms, beginning a chain reaction that produces heat.

The uranium used, in this case, comes in pellets that are less than an inch long. According to Duke Energy, "A single pellet produces the energy equivalent to a ton of coal." Still, many people remain skeptical about the overall benefits of nuclear power, in part because they fear another Three Mile Island.

In February 2007, the Nuclear Regulatory Commission (NRC) issued a summary of events, describing the 1979 accident at Three Mile Island. According to the NRC, "In a worst-case accident, the melting of nuclear fuel would lead to a breach of the walls of the containment building and release massive quantities of radiation to the environment. But this did not occur as a result of the Three Mile Island accident." Instead, immediate steps were taken to gain control of the reactor and ensure adequate cooling. No deaths or injuries to plant workers or to nearby community members occurred. And, ultimately, there were "sweeping changes involving emergency response planning, reactor operator training, human factors engineering, radiation protection and many other areas of nuclear power plant operations."

In 2006, Patrick Moore, cofounder of Greenpeace and an unlikely nuclear proponent, wrote a case for "going nuclear" in which he called Three Mile Island "a success story: The concrete containment structure did just what it was designed to do—prevent radiation from escaping into the environment. And although the reactor itself was crippled, there was no injury or death among nuclear workers or nearby residents."

According to Moore, while fifty-six deaths can be "directly attributed" to the accident at Chernobyl, "No one has died of a radiation-related accident in the history of the United States civilian nuclear reactor program." Compared to the "more than 5,000 coal-mining deaths that occur worldwide every year," said Moore, nuclear energy is one of the safest—and least expensive—forms of energy production we have available.

South of the San Onofre Nuclear Station, along southern California's I-5 freeway, the scientists and engineers at General Atomics are working on a "naturally safe" nuclear reactor known as the Gas Turbine-Modular Helium Reactor (GT-MHR). According to Doug Fouquet, spokesman for General Atomics, "The water reactors in the United States rely on an emergency core-cooling which has to be prompt-acting, and the helium reactor—using graphite fuel and helium as a coolant instead of water—provides an extra margin of safety due to the natural characteristics of the plants...we call it passive safety, not requiring active safety."

The GT-MHR was originally developed as a joint General Atomics and Russian-funded program in order to dispose of reclaimed plutonium from Russian nuclear weapons. In 2001, General Atomics claimed the GT-MHR could also be used for commercial power generation, providing 50 percent higher efficiency than today's reactors, and it was an "inherently safe design" that did not rely on active systems, such as pumps and valves, to maintain safety.

Dr. Alan Baxter, one of the GT-MHR developers, says the GT-MHR has the potential to produce electricity for homes and also to power chemical plants, steel mills, aluminum smelters, and to extract oil from oil shale. In addition, the GT-MHR can be used to produce hydrogen to run automobiles or to liquefy coal to make gasoline with less CO_2 emissions. Says Baxter, "Nuclear power is here to stay. There are currently 440 operating reactors around the world [and] they produce 20 percent or more of the world's electricity without any greenhouse gas emission."

In May 2005, Gallup published findings that 54 percent of the Americans were in favor of using nuclear power as a way to provide electricity for the United States; however, 63 percent said they were opposed to the construction of a nuclear energy plant in their area.

One reason Americans are uncomfortable with nuclear power is because of the radioactive waste that is a by-product of nuclear power production. Opponents argue that the costs of properly transporting, storing, and monitoring spent radioactive fuel are significant and the dangers of improper transportation, storage, or monitoring could be devastating.

The DOE's Office of Civilian Radioactive Waste Management (OCRWM) has been studying a potential site for storing spent fuel since 1978. Yucca Mountain, Nevada, is approximately 100 miles northwest of Las Vegas, on federally protected land in the Nevada desert. The site was originally selected because there were no nearby human populations, no natural resources of known value, a stable geography, and a dry environment that would be conducive to waste storage. However, the project has become mired in political controversy. Currently, according to the OCRWM, "The DOE is in the process of preparing an application to obtain the Nuclear Regulatory Commission license to proceed with construction of the repository."

"In the meantime," says Baxter, "spent nuclear fuel is stored on-site at operating reactors around the country in specially designed, shielded, concrete buildings. Everyone agrees that this is not a long-term, viable solution to the problem, but we lack the political will to go ahead and open Yucca Mountain."

Another alternative for dealing with spent fuel is recycling. Says Baxter, "The original intention of the United States nuclear program was to recycle the spent fuel from the discharged nuclear waste from reactors to extract the material that could be reused, extend resources, and minimize the waste that had to be disposed of." However, he says, "The Carter administration stopped recycling in the United States because it was felt that one of the useful by-products, plutonium, could be diverted to make nuclear weapons. That administration assumed that its policy would be mimicked by other nations around the world." It was not. Says Baxter, "The Carter policy has been ignored by every other nuclear-capable country around the world because it is counter-productive. It maximizes the amount of radioactive nuclear waste that has to be disposed of, while minimizing the amount of energy that can be extracted from uranium."

Global Warming I$ Good for Business

Although there are currently no reprocessing programs in the United States, Baxter claims, "It is possible, by repeated recycling of the discharged fuel, and reuse in specially designed reactors, to eliminate essentially 100 percent of the waste. The key question is economics (i.e., what is the cost to do this versus direct disposal in a high-level waste disposal facility?)"

Says Baxter, "Only 1 percent of the original discharge from [a] reactor is what is called 'high-level waste' or 'transuranics.'" He adds, "The GT-MHR can use these transuranics as a nuclear fuel to produce energy—we have a patent on this process, which we call the 'deep burn' reactor." Baxter claims that, if it were allowed, such a process would further reduce the waste that is finally sent for storage to a Yucca Mountain-type repository.

However, not everyone agrees that reprocessing spent fuel is the best way to go. In the April 2008 issue of *Scientific American*, theoretical physicist Frank N. von Hippel, codirector of Princeton University's Program in Science and Global Security, wrote, "Reprocessing is an expensive and dangerous road to take....Recycling the plutonium reduces the waste problem only minimally. Most important, the separated plutonium can readily serve to make nuclear bombs if it gets into the wrong hands." Von Hippel felt that, while the dilemma over where to store used reactor fuel was real, it was better to store spent fuel in dry casks at reactor sites than to "rush into an expensive and potentially catastrophic undertaking."

Baxter notes, "It is ironic that one of the impurities in coal can be uranium (at the parts-per-million level). However, a large coal fired plant uses so many millions of tons of coal per year to generate electric power that it can discharge to the atmosphere more uranium than would be required to generate the same amount of electricity using a nuclear reactor." He asks, "How do we intend to safely dispose of the millions of tons of poisonous waste from coal-fired plants?"

While some are arguing the pros and cons of nuclear fission, others are looking to the stars at another type of nuclear energy. Fusion has been a reality in the universe for billions of years. The sun is a huge nuclear reactor, which, by some estimates, produces 100 million times as much energy per second as the population of Earth uses per year.

"The fusion process in stars is responsible for element-building as well as energy production," says Rick Lee, DIII-D Tokamak operations fusion education manager at General Atomics. "Many believe it will produce beneficial results for humankind and the environment for millennia."

Fusion is basically a reaction in which hydrogen atoms combine or fuse together to form a helium atom. In the process, some of the hydrogen mass is converted to energy. Proponents of fusion say that if human beings could bring the power of the universe down to earth, our energy worries would basically be over. That is a big "if" according to skeptics who point out that fusion power is an elusive concept. Some of the best minds in science have tried to harness its power for decades and yet no one has been able to produce a functioning electricity-generating reactor to date.

However, as General Atomics' Doug Fouquet points out, "It's too good not to pursue for electric power generation because the fuel would be deuterium [an isotope of hydrogen] that you've got available in ordinary water." In fact, according to General Atomics, one out of every 6,500 atoms of hydrogen in ordinary water is deuterium, which means a gallon of water has the energy content of 300 gallons of gasoline.

Indeed, in Seedmagazine.com's June 2006 article "The Future of Fusion," author Britt Petersen cited the virtually limitless (and inexpensive) supply of sea water available for fuel as one of the most compelling arguments for fusion power. Furthermore, the radioactive by-products from fusion "decay within 50 to 100 years," versus "hundreds of thousands" for some fission by-products. Also, according to Petersen, there could be no Chernobyl with fusion power because the process itself is "so delicate that any error in operation would end the process rather than causing a meltdown."

Currently, there are two fusion reactors under construction. The Lawrence Livermore National Laboratory's National Ignition

Facility, which is due to be completed in 2009 in northern California, will use high-powered lasers to achieve fusion. The International Thermonuclear Energy Reactor (ITER) in France will use magnetic fields to generate the necessary energy to achieve fusion. ITER partners include the European Union, Russia, Japan, China, India, Korea, and the USA. However, U.S. funding for ITER has been problematic since Congress reportedly cut its 2008 budget for the project by over $100 million.

There are many who feel that nuclear power in general—and fusion in particular—is a waste of resources that would be better spent on renewable energy sources such as solar and wind. Yet, the lure of fusion is inescapable. If you look at the history of flight—or, more to the point, look at the generations it took for human beings to evolve from the dream of flying to the reality of actually getting a plane off the ground—fifty years of trial and error is a mere blink of the eye. Proponents of nuclear fusion believe it is only a matter of time before we unlock the secret of sustainable fusion power. When that happens, they say, it will revolutionize the way we live.

9

The Grandchildren's Test

Jim Rogers has a standard by which he measures his biggest decisions—he calls it "the grandchildren's test." According to the chairman, president, and CEO of Duke Energy, "When I am faced with a really big question—and I am faced with big questions, when you think about building nuclear plants and coal plants, where you invest your money and how you maintain electricity that transforms the lives of millions of people every day—I ask myself these questions: Years from now, when my grandchildren look back at what I did today, will they know I did the right thing? Will they look back with pride on the decisions that I make? Will the actions I am taking today create a better life and a brighter future for them tomorrow? I call that the grandchildren's test."

Right now, many of us are facing our own "grandchildren's test" in considering what kind of energy we will use to power our homes and businesses and how we will reduce our own carbon footprint. And, though we may not have the individual impact of a large corporation, our collective decisions will have tremendous repercussions for our planet. Each of us must weigh the trade-offs of various energy strategies in terms of cost, availability, and (if we are wise) sustainability. Right now, both long-term and interim solutions are being suggested which may help us in our quest for environmentally safe and economically sound energy sources. So-called "clean coal" is one idea that has been brought to the table.

The United States has been called the Saudi Arabia of coal. The Energy Information Administration (EIA) claims the United States contains 27 percent of the world's recoverable coal reserves. The World Coal Institute estimates that, as of 2006, we are the second hard-coal producer in the world (China is number one). On average in the United States, coal accounts for almost 50 percent of total electricity generation, with some areas in the Midwest accounting for up to 94 percent of total generation, according to the EIA.

Unfortunately, while burning homegrown coal to produce power for our cities may address the issue of energy independence, it causes a host of environmental and health problems, not the least of which is the production of greenhouse gas emissions. Emissions from coal plants reportedly account for one-third of all human-released mercury and CO_2 emissions in the United States.

The coal that we (and others) burn can have a negative global effect, with pollution from China's coal-fired power plants reportedly increasing cloud cover, raising ozone levels, and increasing health problems in places as far away as the western United States. If we can find a way to reduce coal emissions or capture and safely store them where they cannot escape into the atmosphere, proponents claim we will have a huge reserve of clean and safe domestic energy to feed our growing need for fuel in the twenty-first century, with little or no negative environmental effects. In 2003, the DOE formed, in partnership with the FutureGen Alliance, a non-profit consortium of international coal industry leaders to develop advanced coal-generation and large-scale sequestration technologies. The FutureGen Initiative was supposed to build the first integrated sequestration and hydrogen production research power plant in the world to reduce greenhouse gas emissions through the use of clean coal technologies and also to capture and store (or sequester) CO_2 emissions.

The FutureGen Alliance claimed the facility would be the first of its kind "to combine and test several cutting-edge technologies in a single plant, including coal gasification, emissions controls, hydrogen production, electricity generation, and carbon dioxide capture and storage." By the end of 2007, the Alliance had selected Mattoon, Illinois, as the site for its massive undertaking, and the project was scheduled to begin plant construction in 2009 and go online with operations in 2012. The stakes were high. If successful, the project would create the world's first near-zero emissions fossil fuel plant. However, the costs to build the plant—initially estimated at $1.5 billion—began to climb.

In January 2008, in what *The Wall Street Journal* called "a major policy reversal," the Bush administration suddenly announced that it was no longer going to focus on the Mattoon site but was instead going to implement a "restructured FutureGen approach to demonstrate carbon capture and storage technology at multiple clean coal plants." The restructured approach would focus on separating carbon dioxide for sequestration but would not include hydrogen production as was originally planned.

After initially expressing surprise at the announcement, the FutureGen Alliance reaffirmed its commitment to the Mattoon site and to the original FutureGen vision of a carbon dioxide capture-and-storage and hydrogen production research facility. "This project is further along than any other such project in the world to integrate IGCC [integrated gasification combined cycle] with 90 percent carbon capture and sequestration in deep saline geologic formations, while generating electricity using a first-of-a-kind hydrogen turbine," claimed the Alliance.

In many ways, FutureGen exemplifies the volatile developments that are affecting clean coal today. How—or even if—the FutureGen project progresses is still up for debate. What is not up for debate, however, is the need to do something to reduce and/or capture the amount of greenhouse gases that are emitted from our coal-fired power plants.

Currently, various capture technologies are being researched and developed both by government and private industry. Pre-combustion CO_2 capture and post-combustion CO_2 capture are two areas currently under study.

With pre-combustion CO_2 capture, fuel such as coal is "gasified" or turned into a gaseous form, known as syngas. The CO_2 is captured and sequestered, and the hydrogen-rich syngas is used to power a combustion turbine to generate electricity (and it may also be used to power future fuel cells.) Integrated gasification combined cycle technology allows the steam that is heated by the combustion turbine exhaust to be used to drive a second turbine and generate additional electricity, thus, in effect, producing more power with the same amount of coal.

With post-combustion capture, CO_2 is captured and sequestered after coal is burned to generate electricity. Although post-combustion capture skips the coal-to-gas step needed in gasification, it is technically more difficult to extract the diffuse greenhouse gas after burning the coal. It is also more expensive to capture CO_2 post-combustion from a pulverized coal plant, according to Michael G. Morris, chairman, president, and CEO of American Electric Power (AEP).

Today, AEP is just one of several companies today that are looking into the feasibility of constructing clean coal power plants, utilizing IGCC technology. According to Morris, "With restrictions on carbon dioxide emissions expected in the future, IGCC technology represents an important advancement for power generation and for the coal industry." Pending permitting and approval from various agencies, AEP has plans to construct a 629-megawatt IGCC plant in West Virginia. The company has also proposed another IGCC plant in Ohio.

Duke Energy is moving forward with the production of an IGCC coal power plant as well—this one located in Edwardsport, Indiana. At the beginning of 2008, the company got the go-ahead to begin construction on the plant, which is due to be completed by 2012. According to Duke, "The current 160-MW [pulverized coal] plant at Edwardsport emits approximately 13,000 tons of sulfur dioxide, nitrogen oxide, and particulate emissions annually. It operates about 30 percent of the time. Preliminary data indicates a 630-MW IGCC plant operating 100 percent of the time will emit about 2,900 tons of those same pollutants (including mercury) annually."

In addition, the company claimed, "The proposed IGCC plant will use an average of 11 million gallons of water per day as compared to the current plant that uses 188 million gallons per day." Although the plant does not utilize carbon capture and sequestration capabilities, Duke has "proposed studying the capture and sequestration of a portion of the plant's carbon dioxide emissions."

General Electric is a leader in IGCC technology and claims that "The chemical industry began using gasification to make chemicals such as ammonia and fertilizers [in 1950]. However, the feedstock was mostly crude oils rather than coal." Since 2006, BP has been working with industrial partners such as GE and with Edison Mission Group to develop hydrogen power by gasifying petroleum coke, a derivative of oil refinery coker units, at its Carson, California, refinery.

In a 2006 speech at the Los Angeles World Affairs Council, BP's chief executive of Gas, Power & Renewables, Vivienne Cox, said, "In our new power plant, the coke and treated wastewater will be converted to hydrogen and CO_2. The hydrogen would then be used to fuel a gas turbine to generate electricity—enough for more than 300,000 homes. Meanwhile the CO_2 would be captured and transported through a pipeline to an existing oilfield where it would be injected into reservoir rock formations. This stimulates extra oil production and permanently traps the CO_2. So it's a triple win. California would acquire 500 MW of new, secure, generating capacity. Four million tons of CO_2 a year would be eliminated from the atmosphere. And some aging domestic oil fields would be given a new lease on life."

Proponents see great promise in clean coal technologies. However, many environmentalists and others believe so-called "clean coal" is nothing more than wishful thinking at best and "a blatant lie" at worst. In a July 2006 Treehugger.com article entitled "Important! Why Carbon Sequestration Won't Save Us," science and technology editor Michael Graham Richard outlined several problems, not least of which was that new clean coal plants are prohibitively expensive to build and operate and would take decades to make any significant contribution towards reducing greenhouse gas emissions.

Referring to Tim Flannery's book *The Weathermakers*, Richard went on to say that CO_2 must be compressed into liquid in order to be injected into the ground—"a step that typically consumes 20 percent of the energy yielded by burning coal in the first place." In addition, there are not enough A-grade reservoirs near power stations to accommodate large-scale geosequestration. And using substandard reservoirs or failing to adequately monitor reservoirs "from that day on" in perpetuity could result in harm and even death to local populations if CO_2 leaks occur.

In short, Richard argued, the costs are not worth the benefits when it comes to gambling on inefficient clean coal technologies that use about 25 percent of the energy they make just to keep operating. He believes it would be better to spend our time and resources pursuing truly clean energies such as solar, wind, wave, or geothermal than to waste it on a technology that simply is not feasible.

As contentious as the debate over coal power may be, the reality is, given the amount of coal and the growing demand for power in this country, some trade-offs will most likely be made. Whether those trade-offs and the interim solutions that evolve from them are environmentally sound, economically feasible, politically expedient, or just plain random depends upon the choices we make.

10

The Earth's Heat

Dr. Eylon Shalev is on geologic time—he measures the world in terms of millennia, not years. Because of his background in geology, he says he has a different way of looking at things: "While I think that global warming is real and it's actually happening, I don't think that it is as dire as some people say for the simple reason that the main cause of global warming is the CO_2 that we humans are releasing into the atmosphere….But the fact is that we are running out of the thing that produces carbon dioxide. We are running out of hydrocarbons."

That is not to say we should stop trying to develop renewable sources of energy. Says Shalev, "If we can produce energy without releasing CO_2 in the atmosphere, that would definitely be helpful." For him, the most promising source of renewable energy today is geothermal—an energy source that is as old as the Earth itself.

According to the Department of Energy, "Human beings have used geothermal energy in North America for at least 10,000 years." In the United States, Paleo-Indians used hot springs for warmth, cleansing, healing, and relaxation. And, today, engineers are "developing technologies that will allow us to probe more than ten miles below the Earth's surface in search of geothermal energy."

The word "geothermal" means "Earth's heat." Basically, magma from the interior of the earth heats reservoirs of water that have formed in the fractured and porous rocks underground, and this hot water, or geothermal fluid, can be used to produce thermal and electrical energy.

Global Warming I$ Good for Business

The Geothermal Energy Association lists three different ways to use geothermal energy: Direct use does not require geothermal energy to be converted into electricity before it is used. Ancient Roman baths and Native-American hot springs are examples of direct use of geothermal applications, which have been used for centuries. Geothermal heat pumps are another way to use geothermal energy. Geothermal heat pumps are devices that are installed on-site to heat or cool residential and commercial buildings, utilizing the Earth's relatively constant interior temperature. The third use for geothermal energy, known as geothermal electricity, provides clean, reliable energy to the electric grid.

Geothermal electricity is being generated in several places in the United States that have geothermal resources. The Geothermal Energy Association claims that, as of 2003, geothermal is the third largest source of renewable energy in this country. Furthermore, the United States produces 32 percent of the world's total—more geothermal electricity than any other country. The largest complex of geothermal power plants in the world is the Geysers in Northern California, according to Calpine Corporation, which owns and operates nineteen of the twenty-two power plants in operation there. With a net generating capacity of approximately 725 megawatts of electricity, the Geysers is said to produce enough energy to power 725,000 homes, or a city the size of San Francisco.

In geothermal electricity production, wells are drilled into geothermal reservoirs, which contain hot water or steam. These wells pump the geothermal fluid to the surface where it can be converted into electricity. In the case of the Geysers, wells as deep as two miles are drilled through cap rock to tap natural steam, which is in turn piped to the generating units in the Geysers complex.

Global energy giant Chevron Corporation claims to be the largest private producer of geothermal energy in the world, accounting for more than half of all privately developed geothermal power. However, according to Chevron, there are relatively few places in the world where molten rock can surge up through fractures in the Earth's crust near enough to the surface (usually around 3,000 meters or 9,800 feet) to heat underground water. At these places, the company drills wells "similar to those used to produce oil and natural gas" to recover hot water and steam from geothermal sites and convert it into energy.

Puna Geothermal Venture, located on the big island of Hawaii near the Kilauea Volcano, is owned by Ormat Technologies, Inc., and has been in operation for fifteen years. Puna Geothermal Venture currently replaces about 144 thousand barrels of oil annually, according to plant manager Mike Kaleikini. "We're just like any other large generating facility. The only difference is that the resource we use is the geothermal steam."

Kaleikini estimates that approximately 20 percent of the annual energy consumed on the big island of Hawaii comes from Puna Geothermal Venture. However, he says, the site could produce more energy if the facility expanded its operation and drilled more wells. There is very little incentive to do that, though, since the island's total available energy capacity exceeds its residents' current needs. Unless demand increases on the big island, fossil fuel plants are decommissioned, or transmission lines to other islands are put in place, Puna Geothermal Venture will probably continue to supply the same amount of energy it has since the mid-1990s. However, according to Kaleikini, "If there was ever an opportunity [to expand] I think we would be very much interested."

Like many geothermal facilities, Puna Geothermal Venture is a closed-loop system. The hot water is extracted from the ground and the steam from that is used to produce energy before it is condensed back to water and then rerouted underground into injection wells. "Everything that comes up eventually goes back down," says Kaleikini. "So we don't have a plume or anything like that. We have air-cooled condensers; so, as far as killing ozone and that kind of thing, no, we're definitely not doing any of that." However, Kaleikini adds, there is "vog" or "volcano smog" that comes from the volcano itself at certain times that is a natural occurrence, unrelated to human activities.

The Seismology Group, formerly located at Duke University, has been studying geothermal energy for several years and has since moved on to the University of Auckland in New Zealand to continue its work. Dr. Eylon Shalev has worked with the group, measuring seismic activity in many places, including Hawaii's Puna Geothermal Venture.

"The biggest problem with geothermal energy is that it is easy to produce and economical to produce at a certain spot on the face of the earth; and those spots are where you have geothermal fluid at high temperature close to the surface of the earth," says Shalev. "The problem is that all those places on the earth—or most of the major ones—have already been tapped or are in the process of development. If you go to any other place on the earth, in order to get the right temperature for geothermal, you need to drill not to a depth of one or two kilometers but rather to a depth of at least six to eight kilometers. And, since drilling is extremely expensive, it just makes it uneconomical."

The Seismology Group, led by Dr. Peter Malin, works to enhance or engineer geothermal conditions in places where they might not otherwise be present. They utilized a type of enhanced or engineered geothermal systems technology known as "hot dry rock" in Basel, Switzerland. Says Shalev, "If you want to produce energy in such a way, the plan includes drilling to a depth where it is hot enough—and in Basel it was five kilometers—pumping cold water through cracks in the hot the rock, thus heating the water, and then extracting the now heated water through another well. The end result is that you have a closed circuit; you inject cold water, extract hot water, use the heat, and then you re-inject the water." The problem, according to Shalev, is that there are few natural cracks in the rocks at such depth: "So the first step required is to pump water under great pressure into the well in order to actually crack the rock, thus enhancing the geothermal system."

Unfortunately, in Basel, the "cracking of the rock" also produced seismic activity which, while not unexpected, produced a series of mild earthquakes (including one 3.4 magnitude quake) which did little damage but shook up the local population and caused the project to be put on hold. Shalev admits, "When you inject water to crack the rock, you are essentially producing earthquakes. Now, usually they are very small earthquakes and earthquakes that people can't feel because they are below the threshold. They are perhaps magnitude one or one-and-a-half, and people can't feel them. But, when you try to crack the rock you can produce a larger event, and that's actually what happened in Basel." Still, Shalev insists that, with proper monitoring, the seismic activity is not a serious long-term environmental issue because the rock-cracking stage is limited in time.

According to a 2007 geothermal market report put out by the Global Sustainable Energy Team of Glitner Bank, a leader in geothermal energy financing, not only is geothermal energy both clean and renewable (its by-products consist mainly of water, which can be re-injected into the ground), it "represents the only real baseload capacity alternative to fossil fuels, such as coal or oil." Furthermore, with new technological developments such as enhanced or engineered geothermal systems, geothermal energy could be developed anywhere in the United States, instead of merely in a few of the Western states.

A 2006 Massachusetts Institute of Technology report also endorsed geothermal energy production in general and enhanced geothermal systems in particular. According to the report, "Many attributes of geothermal energy, namely its widespread distribution, baseload dispatchability without storage [which is not available with renewables such as solar and wind], small footprint, and low emissions, are desirable for reaching a sustainable energy future for the United States." The report discusses the benefits of continued research, development, and demonstration in several areas, including the use of CO_2 as a reservoir heat transfer fluid. Utilizing CO_2 in enhanced geothermal systems could both improve the performance of the reservoirs themselves and provide a means of sequestering carbon in stable underground formations.

Global Warming I$ Good for Business

In the United States, enhanced geothermal systems technology is currently being commercially developed at Ormat Technologies' Desert Peak site outside of Reno, Nevada, in order to enhance the production of the existing geothermal reservoir there. In a 2008, Ormat chairman and CTO Lucien Bronicki announced, "Our objective in the Desert Peak [enhanced geothermal systems] project is to demonstrate that [enhanced geothermal systems] technology can achieve its potential of providing 100,000 MW of clean, baseload power...and show that this technology will enable geothermal electricity to be produced in regions where it is not currently economically viable." If, as Ormat suggests, 1 MW can provide power for 750 homes and offset 7,500 tons of CO_2, then 100,000 MW could make a substantial impact on energy independence and greenhouse gas emissions in this country.

Many scientists, including Shalev, believe that geothermal energy may be the long-term solution to sustainable low carbon energy independence in the United States. "Humans started using hydrocarbons on a large scale about 100 years ago, and we have already used about one-third of it," says Shalev. "When you think of it, the era of using coal or oil or whatever hydrocarbon-based resource is a very short one compared to even recent human history. So, when you try to think what comes after this era, I think you have only three viable answers: One is solar energy, one is nuclear energy, and one is geothermal energy." Of the three, he says, "I think that geothermal has the chance to be the most important energy source in the future."

Says Shalev, "While you and I probably won't be alive when we humans run out of hydrocarbons to burn, our grandkids will. It is very simplistic and egotistical to say that we live only for the present and future generations will take care of their own problems." For now, he believes the most effective thing human beings can do is conserve "because a new type of energy system is hard to develop and is very, very costly." Conservation "gives us more time to try to find another energy solution for the planet."

The Earth's Heat

Many individual American home and business owners believe they already have found another energy solution—geothermal or ground source heat pumps. According to the International Ground Source Heat Pump Association, "Ground source heat pumps are electrically powered systems that tap the stored energy of the greatest solar collector in existence—the Earth."

With ground source heat pump systems, polyethylene piping is placed underground, where the earth maintains a relatively constant temperature of anywhere from forty to seventy degrees Fahrenheit, depending on latitude. In a closed-loop system, a water solution is circulated through the pipes, carrying heat from the earth into the building during winter and pulling heat from the building and dissipating it underground during summer. In an open-loop well system, water from a well is pumped into the building and then discharged outside. While this installation is usually less expensive to install than a closed-loop system, there may be local codes regarding discharging methods. Neither the open nor the closed-loop system uses an outside compressor unit (such as the air conditioning units most Americans have behind their houses). There is only an indoor unit, which is approximately the same size as a washing machine and is usually placed in an inconspicuous place in a basement or attic.

The Geothermal Heat Pump Consortium, a nonprofit organization, claims that a 2,000-square-foot home can be heated and cooled for about a dollar a day. And installing just 400,000 geothermal heat pump or geoexchange systems each year could cut greenhouse gas emissions in the United States by one million metric tons of carbon, the equivalent to planting over a million acres of trees or converting over half a million cars to zero-emission vehicles.

According to John Kelly, executive director of the consortium, geothermal heat pumps now account for approximately one percent of the United States heating, ventilating, and air-conditioning market, with "double-digit" annual growth in recent years. Estimates are that 60 percent of the residential units sold are for retrofits for existing homes and around 40 percent for new home markets. The

81

commercial market is picking up as well and accounts for half of the overall geothermal heat pump usage. While most shipments for geoexchange units have typically gone to the South and the Midwest, Kelly says there is now quite a bit of growth in the Northeast and the Western states as well.

"People are recognizing the advantages," according to Kelly. "And the impact of that fact is being felt throughout the [heating, ventilating, and air conditioning] industry....For contractors, the opportunities are significant. Consumer demand for cost-efficient, 'green' energy systems is putting pressure on builders. Builders, in turn, are consulting contractors about how best to proceed. And that gives contractors a chance to differentiate themselves."

Enertech Manufacturing, makers of Energy Star-rated Geo-Comfort geothermal systems, touts many advantages to using geo-exchange systems, including annual savings of 30 to 40 percent compared with ordinary systems, a positive cash flow in three to five years, a twenty year lifespan for its system, and environmentally safe operations. In addition, geoexchange systems do not produce any carbon monoxide, which traditional furnaces can, and they are quiet, unlike most outside air conditioning units. They also heat water and dehumidify the air within the home or office.

In fact, according to the EPA, geothermal heat pumps are "the most energy-efficient, environmentally clean, and cost-effective systems for temperature control," which may be why proponents say that, for individual homeowners and businesses looking for relief from high gas and oil prices, geothermal heat pumps are the best thing under the sun.

11

Uncharted Waters

The Grand Coulee Dam is the largest producer of hydroelectric power in the United States and third largest in the world. A book by Murray Morgan, *The Dam*, captures the spirit of the men who built some of the most spectacular renewable energy facilities this country has ever seen: "There are parts of our culture that stink with phoniness. But we can do some wonderful things too. That dam is one of them. If our generation has anything good to offer history, it's that dam. Why, the thing is going to be completely useful. It's going to be a working pyramid. I just want to help build it."

Hydropower has been an energy source for mankind since the Greeks turned waterwheels to grind wheat over 2,000 years ago. The power of water has inspired everyone from the engineers who designed the great dams of the early twentieth century to the common men who risked their lives building something "completely useful" to power this country in perpetuity. At its peak, hydropower provided 40 percent of electrical generation in the United States (today that figure is closer to 10 percent).

The DOE lists three types of hydropower facilities in operation today. Impoundment facilities, which are the most common, dam water in a reservoir, from which it can be released to flow through a turbine, thus activating an electrical generator. Pumped storage facilities pump water from a lower reservoir to an upper reservoir during low demand and then release the water back to the lower reservoir to generate electricity during high demand. Diversion, or run-of-river, facilities divert the water through a canal or penstock with or without a dam.

Hydropower facilities in the United States can generate enough power to supply 28 million households, equivalent to almost 500 million barrels of oil, according to the DOE. However, while many large facilities such as the Grand Coulee Dam generate renewable energy, their environmental sustainability has come into question by many who are concerned with the impact they have on the surrounding and downstream natural ecosystems. Entrepreneurs and others are now looking at small or even micro hydropower plants to provide electricity in a more sustainable way.

The DOE defines large hydropower as facilities that have more than 30-MW capacity (Grand Coulee has a capacity of 6,480 MW, just to give you an idea of the scope). Small hydropower facilities have a capacity of 100 kW to 30 MW, and micro hydropower plants produce up to 100 kW, which is enough electricity to run a home, farm, or small village.

Trey Taylor, president of New York-based Verdant Power, hopes to capture the power of hydro's renewable energy in an environmentally friendly way. "I saw a huge market need and a problem at the same time," says Taylor. "Could we design technology that could harness water currents and convert that to electricity too, with the least amount of environmental impact?" In fact, that is exactly what they have set out to do.

Verdant Power's systems use no dams, canals, or penstocks. Instead, they utilize unconventional kinetic hydropower systems that Taylor says are more akin to wind power than hydropower. With wind power, says Taylor, "it's the sweep of the blade...and it's the speed of the current that determines how much power you can produce in each one of the turbines, and our systems work the exact same way." Verdant Power's three bladed downstream systems, which actually look like underwater windmills, have rotors at the rear of the turbine (versus the frontward-facing, upstream rotors on wind farms); this configuration allows them to change direction with the flow of the water in either tidal or river settings.

Verdant Power says its Roosevelt Island Tidal Energy Project in New York City is the world's first multiunit, grid-connected kinetic hydropower field. Not only has it produced more than 45 MW of power but, says Taylor, that power is actually delivered to customers in Manhattan (including many of the electric-powered park maintenance vehicles on Roosevelt Island). He hopes, someday, as many as 300 of the 5-meter (16-foot) turbines will be deployed in the area. The Roosevelt Island system is specifically designed for the depth and speed of the East River, says Taylor. Projects in other places may have different sized generators, depending on the tidal or river conditions.

Verdant Power's second project, the Cornwall Ontario River Energy Project in Ontario, Canada, will have much larger generators because the St. Lawrence River runs faster than the East River. Also, the St. Lawrence is not a tidal channel, which means it is flowing twenty-four hours a day and has a higher capacity factor than the East River project. A power plant's capacity factor is determined by dividing the actual power the plant produced over a given period of time by the amount of power it could have produced had it run at full capacity during that same time period. According to Taylor, tidal power has a capacity factor of approximately 30 to 32 percent, about the same as wind power on land.

Although similar in some respects, tidal differs from wind in that, while it is an intermittent source of power (coming on and going off as the tides come in and go out), tidal power is reliable and predictable because we know the tides will come in and go out and we can predict when that will happen. Wind, on the other hand, tends to be more episodic. Taylor says tidal power allows a utility such as Con Edison to "back their power off or bring it on very seamlessly." The end result is that consumers won't even realize a portion of their power is coming from the tides. River power, he claims, has an even higher capacity factor—up to 80 to 90 percent—because it runs twenty-four hours a day, which would make it more suitable as a baseload power source.

Because it is such a new technology, Verdant Power has had to spend years gathering evidence and researching the environmental impacts of its systems in order to get the necessary permits to move forward with commercial production. Says Taylor, "It's going to be a lot faster in the future than right now because we are the first ones in the world to sort of run the gauntlet." So far, claims Taylor, Verdant Power is the only company to build and field test four different types of kinetic hydropower turbines. The company has tested horizontal axis turbines, which are similar to wind turbines, as well as vertical axis turbines, which are similar in appearance to Gorlov Helical turbines and "look more like an eggbeater or an industrial Mixmaster," according to Taylor.

Another of Verdant Power's projects will utilize the vertical axis turbine at a Dow Chemical industrial canal in Texas. Says Taylor, "If it works, what we could do is end up providing distributed generation for a plant right on-site, so it's bypassed the transmission grid altogether." Taylor hopes the Aqueduct and Canal Energy Project, as it is called, will not only convert the shallower, slower moving canal water into electricity but will also pump water through electrified membranes to produce clean water.

"Dow is the world's largest manufacturer of reverse osmosis membranes," says Taylor," and so what intrigues us about this kind of relationship...is [the chance] to create integrated water and clean energy systems....Water is now more expensive than oil, but we can live without oil, but we can't without water. That's why we want to get into that sector and especially if we can create modular units that would easily deploy using local resources."

Taylor sees a huge potential for energy capture utilizing this vertical axis technology, not only in industrial canals and aqueducts but in water treatment facilities and bypass channels: "In the United States, there's over 70,000 dams and only 2,300 produce hydroelectric power. All the rest of the dams are used for flood control or irrigation or waterbed, but they're always releasing water or bypass channels; so rather than letting that energy go, capture it and make electricity while sending it on its way."

Verdant Power's vertical axis Rapid Flow System is designed to work in waters that are moving less than one meter per second and in water that is six feet deep (compared to the horizontal axis Free

Flow System which is cost-effective in water that is moving two or more meters per second at a depth of thirty or more feet).

Ironically, though, Taylor says that the biggest obstacle his company has had to overcome is not technological but the "regulatory process in the United States." Although they work closely with the Federal Energy Regulatory Commission, which is "doing all kinds of things to help speed up the process and move us along," Taylor says the resource agencies in the United States are based on adversarial processes designed to enforce the 30-year-old Environmental Protection Act: "Everybody's afraid of being sued by one another and have to keep asking for one study upon another, but the cost for these studies fall to a start-up company, not the government.... Right now, we're spending more money on studies than we are on the design, building, and implementation of the technology, and it just hurts, and it slows us way down...." Canada and the United Kingdom, by comparison, have developed a more collaborative process that "makes a huge difference in helping do things better, trying to work toward the better good."

"In the United Kingdom, there used to be the Department of Trade and Industry. The new department is called Business Enterprise and Regulatory Reform." The idea of government resource agencies helping new industry thrive rather than just protecting the status quo is one which Taylor hopes we can import into this country. "It's a shame," he says. "After all, what gives us our reputation in this country is our ability to innovate."

In the meantime, however, Taylor sees tremendous potential for developing hydroelectric power in developing countries that do not have the traditional transmission grids which waste energy in the transmission process. He envisions small and micro systems, such as Verdant Power's kinetic hydropower systems, providing local power to communities any place in the world.

Ocean Renewable Power Company is also gearing up to extract energy from tidal currents, and they expect to access deepwater ocean currents at some point in the future, as well. According to Chris Sauer, company president and CEO, "There are basically four types of ocean energy: wave energy, tidal energy, deepwater ocean currents, and ocean thermal energy conversion. Sauer's company has targeted tidal energy and, at some point, will also pursue deepwater ocean currents such as the Gulf Stream, which he likens to "rivers within the ocean that flow constantly and tend to be in very deep water."

At the end of April 2008, the company completed initial testing of its proprietary turbine generator unit (TGU), which Sauer explains, "is the heart and soul of our technology." At first glance, the TGU looks a little bit like a combine header (the cylindrical reel that cuts the crops during harvest). However, a typical TGU consists of two horizontal turbines which drive an underwater generator placed between them. Four TGUs can be stacked and mounted into one module, measuring approximately 65-feet long by 65-feet high. At peak, says Sauer, an OCGen module can generate as much as 1 MW of electricity in a 6-knot current beneath the surface of the water.

The depth at which future modules may be placed depends on the site. "Our equipment is deployed below the water, in fact we want to get away from the waves," says Sauer. For example, the company's current site in Western Passage, Maine, along the Canadian border, is in a major shipping channel, so the equipment will be at least sixty-five feet below the water surface. However, at other sites, where there is only recreational boating, the modules can be placed twenty feet below the surface. Also, says Sauer, the TGU can be put on foundations, such as pilings, and used by itself in very shallow estuaries and in river applications. Depending on the location, there could be anywhere from a couple of dozen to hundreds of modules under the water. In a classic tidal application, a power cable would link each of the modules to the main underwater transmission line, which would then transmit the power to shore.

One of the challenges—what Sauer calls "the 800-pound gorilla in the room"—is power transmission: "If you look at how the transmission lines were originally built, first of all they were built by utilities, and they owned their own power plants at that time. So they were essentially built to connect their very large power plants, which were built inland for the most part, to the population centers. If you look at where the best ocean resources are, it doesn't necessarily match up with where the best grid is. So there are going to have to be some investments in transmission infrastructure for this industry to realize its full potential." That potential could include providing 10 to 15 percent of the total electricity needed in the United States, according to Sauer.

"I think, particularly when you look at five years from now, we will be doing projects not just in rivers and in tidal sites...but we'll also be starting to install major projects in deepwater ocean currents," says Sauer. "The exciting thing about that is, even though you are dealing in much deeper water—maybe 1,000 or 1,500 feet of water—you have a constant stream of water going by, and these are huge resources. You really don't have any site constraints in the sense that there is no shoreline nearby. You are going to be twenty to twenty-five miles out in the ocean."

Ocean Renewable Power Company's first commercial project is expected to be put in place by 2010. In 2012, assuming all goes well, Sauer says the company will be in position to start building commercial projects in Maine, Alaska, and elsewhere. In addition to the tidal currents in New England and the deep ocean currents in the Gulf of Mexico, there are more tidal and some ocean current resources from San Francisco to British Columbia and even into Alaska, which also has "great potential" for river applications. Sauer says his company will sell power to large wholesale energy suppliers as well as to the grid, which in New England is operated by the ISO, or Independent System Operator. In Alaska, Ocean Renewable Power Company could sell directly to remote area communities.

For now, however, Ocean Renewable Power Company, like Verdant Power, is navigating uncharted waters as it makes its way through the Federal Energy Regulatory Commission permitting process. Says Sauer, "It was developed basically for large hydro projects. We go through essentially the same process which you would go through if you were going to build the Grand Coulee Dam. Clearly this is different—of course, no one had even an idea of technology like this back when they put these rules together."

According to Sauer, "There is a lot of infighting between [the Federal Energy Regulatory Commission] and [the Minerals Management Service], which is the federal agency that essentially regulates anything that's greater than three miles offshore. So there is a lot of uncertainty there. A lot of risk in terms of permitting [which] could significantly delay implementation of these projects. Even worse than that, it has the potential of driving away investors who might otherwise invest but don't really want to take what they call 'the permitting risk.'"

Much of the problem, says Sauer, is a lack of political will: "If we could get the average citizen to be telling their elected officials, 'Hey look, I want this to be a priority,' that's a very powerful tool. I mean the alternative is—I have to be honest, we are evaluating it—we go to Europe where it's much more favorable. That's sad. I want to do this in the United States. I want to make this a United States industry, but, at the end of the day, if we can't get it done here, we are going to go elsewhere."

12

Here Comes the Sun

In the spring of 2007, Lord John Browne, then BP Group chief executive, stood before a group at Stanford University and said, "The time to consider the policy dimensions of climate change is not when the link between greenhouse gases and climate is conclusively proven but when the possibility cannot be discounted." Coming from one of the leaders in gas and oil development, such a comment was nothing short of extraordinary.

BP—as in "Beyond Petroleum"—has been one of the first major oil companies to jump on the renewable energy bandwagon. The BP logo, Helios, resembles a shining sun and exemplifies the company's interest in and development of renewable energy. Headquartered in the United Kingdom, BP fuels are sold worldwide, including at BP, Amoco, and Arco stations across the United States; its Castrol products are almost synonymous with motorcycle and car racing.

While BP is most definitely still in the oil and gas business, it is also actively working to develop renewable and alternative energies, including solar, wind, biofuels, and hydrogen power with carbon capture and sequestration. With over 2,200 employees worldwide (550 in the United States), BP Solar—headquartered in Frederick, Maryland—is one of the world's largest solar companies and has purportedly brought solar power to more than 400,000 residents in 150 remote villages.

Solar energy is one of the best areas for potential growth, according to Geoff Slevin, who is vice-president of sales and marketing for BP Solar in North America. Slevin points out that many developing nations have bypassed traditional telephone systems, which are heavy on infrastructure, and jumped directly into cellular phones, and he sees the potential for distributed solar energy to evolve in much the same way: "The cost-savings are not having to build large plants, distribution lines, switching stations, and everything that goes on with the large power plant infrastructure—if you jumped from nothing to distributed generation, be that solar or another kind, there are real benefits to people in the developing world."

Half a world away from where Slevin works, Saudi Arabia is exploring renewable energy sources such as solar to power road signs and emergency telephones along remote stretches of desert highways. According to its Royal Embassy, Saudi Arabia receives 105 trillion kilowatt hours of sunlight a day, roughly equivalent to 10 billion barrels of crude oil in terms of energy. This is something even the oil-rich countries of the Middle East are interested in taking advantage of.

Closer to home, BP Solar is looking for ways to make solar more efficient, less costly, and more competitive with conventional energy sources such as coal and oil. "We are after ways to make solar affordable for everyone," says Slevin. Part of this effort involves utilizing better sales and marketing and installation techniques to reduce costs.

One of the main barriers to the use of solar energy is the cost of the initial capital outlay, especially for large commercial systems, which can take decades to pay for themselves.

BP Solar has developed two different methods for large customers to gain the benefits of solar electricity: (1) They can buy the system outright and then own it and all the power it generates, or (2) they can simply allow BP to install the photovoltaic (PV) system on-site and then buy the power directly from BP via a "solar services agreement."

In 2007 and 2008, BP installed PV systems at various Wal-Mart facilities in California at no capital cost to the retail giant. The agreement is that Wal-Mart will buy its electricity from BP for a ten-year period at a specified price in much the same way they would buy energy from an electric utility company. There are a number of benefits for this, according to Slevin, not least of which is that Wal-Mart knows what its PV-generated electricity rates will be over the next ten years, regardless of where traditional utility-generated electricity rates are. This "hedge against rising electricity prices" is attractive to customers who want to lock in savings and reduce their electricity costs.

Says Slevin, "Many companies now are realizing the economic benefits of operating more sustainably, either [because] consumers are willing to pay more for a product if it's created sustainably or just realizing cost-savings to the bottom line by employing sustainable practices like more efficient lighting, temperature controls, and using solar electricity. I don't know what's going to happen to the price of oil, coal, and natural gas over the next decade, but I know the sun's going to come up every day. And, because it's going to come up, I know what the electricity prices for a customer like Wal-Mart are."

But BP Solar does more than make large solar systems for commercial customers in America. Recently, it also partnered with Home Depot in an effort to make solar energy more accessible to the average consumer. In-store displays consist of a video, which explains how PV systems work, as well as a brochure that explains the 1-2-3s of working with a PV system installer. From there, consumers can contact an installation professional for a free in-home consultation to see if solar is right for them.

"I wonder if we as consumers, or as people who operate companies, would think differently about energy if it were colored. So what if energy were blue? Would you insulate your house differently? Would you turn the air conditioner down? It's because emissions and energy losses are things that we can't see usually. I think a lot of times we don't really think about them the way that we should," says Slevin. "One of the biggest hurdles that we have in solar is just education and awareness—of people realizing the fact that they can actually buy this stuff for their own house."

Although there is no denying the initial outlays can be large, Slevin claims "that solar makes remarkable sense for a whole lot of people who don't know that it makes sense for them even now.

And with the continual cost declines and continual improvements in technology, it's going to become more and more available for that next level of implementation." BP is currently working with the California Institute of Technology to create a new type of solar cell from tiny silicon nanorods.

Nanotechnology is probably the single most game-changing technology in the world right now. The Center for Responsible Nanotechnology defines it as "the engineering of functional systems at the molecular scale." The idea is to work with the crystal structures of silicone chips, for example, in order to build an intrinsically more efficient PV cell "from the bottom up" at the nanoscale. To give you an idea how small this is, a human hair is approximately 50,000 nanometers in diameter.

Although nanotechnology is not confined to solar PV production—it is used in everything from clothing to cosmetics—it is certainly causing a great deal of excitement within the solar community. In 2007, researchers at the New Jersey Institute of Technology reportedly discovered a way to combine carbon nanotubes with carbon Buckyballs (also known as fullerenes) to form an organic solar cell that may someday be painted on or printed out using an inkjet printer. Given this technology, homeowners could theoretically print out sheets of solar cells and cover the walls or roofs of their homes to convert sunlight into electricity, and cars could be painted with solar paint in order to power their electric systems using sunlight. The possibilities of such a technology are virtually endless, which is why companies like BP and universities like CalTech are also actively pursuing its research and possible development.

In the meantime, they and others like them are continuing to perfect the traditional PV solar systems, such as the ones that are mounted on rooftops. Of course, since these solar electric systems basically convert the sun's light into electricity, they function best when they are getting full sunlight. A PV cell or solar cell is the basic building block of a PV system. PV cells can be joined together to form PV modules or panels, which can in turn be joined together to form PV arrays, depending on how much energy is needed.

A small solar calculator might only need one solar cell, while a large power plant system, such as the fourteen-MW solar tracking system at Nellis Air Force Base, might consist of thousands of arrays. At Nellis, the PV arrays, or SunPower Trackers as they are called, actually rotate up to twenty-five degrees to follow the path of the sun and to maximize the amount of sunlight captured.

Many see renewable energy production such as solar as more than just a way to reduce our dependence on foreign oil. Rhone Resch, Solar Energy Industries Association president, calls solar energy an "economic engine" for the United States. According to Resch, the recent growth in the solar energy industry infused over $2 billion into the United States economy and created 6,000 jobs. Resch credited United Solar, a thin-film manufacturer, with creating hundreds of jobs in Michigan, a state which has suffered high unemployment. United Solar Ovonic, LLC manufactures UNI-SOLAR products and, with over 1,300 employees, claims to be "the world's leader in thin film solar technologies and the manufacture of thin film solar electric modules and laminates."

Silica, also known as quartz sand, is a common element; however, the process required to render it useful for such things as semiconductors, microchips, and solar cells is expensive, causing many in the industry to look at cheaper materials with which to make solar cells. Thin film solar cells are one of the fastest growing alternatives to crystalline silicon solar cells, and, according to the 2007 Solarbuzz World PV Industry report, "Thin film production more than doubled from 181 MW in 2006 to 400 MW in 2007, accounting for 12 percent of total PV production."

UNI-SOLAR thin film amorphous photovoltaics comes in flexible panels and in rolls of flexible PV laminate (PVL) that are easy to install and remove, require no roof penetrations, and can be used on everything from shade structures to curved metal roofs. Its peel-and-stick backing makes applications both easy and more cost-effective. Although the size and weights of various PVLs vary, none of the laminates use glass, which means the PV system is both substantially lighter and is "virtually unbreakable."

United Solar claims its solar systems are three to five times lighter than conventional modules, yet still provide more kilowatt hours of energy than crystalline panels with the same power rating. The higher output is due to UNI-SOLAR's triple-junction technology, which captures three distinct waves of sunlight and generates more power under low light (i.e. morning and evening) conditions.

In his annual shareholders' report for EDC Ovonics (United Solar Ovonics' parent company), Mark Morelli, president and CEO, demonstrated that solar electricity is approaching "grid parity" with conventional electricity. According to his graphic presentation, the cost per kWh for solar electricity is quickly approaching the point where it will be equal in cost to conventional grid electricity. If the cost of fossil fuels continues to rise, solar energy may actually become cheaper, and many in the industry believe that if the "full cost" of fossil fuels—including the economic, environmental and health costs of producing electricity using those fuels—is factored in, solar energy already is more cost-effective.

While thin film manufacturers such as United Solar rely on products that contain no glass, a newer company, Prism Solar, designs PV systems that utilize the power of sunlight refraction through glass panels to dramatically increase their output. In 2007, the New York start-up was awarded the "Most Promising Technology" award at the Cleantech Network Investment Forum for its unique design.

"The key to our technology is the holographic portion," says company president and CEO Rick Lewandowski. "It splits the spectrum, it splits the sunlight, and we take the portion of the spectrum that we want, send that over to the solar cell…to create what's called TIR (total internal reflection) within the glass—the pane of glass itself—and the glass becomes a light channel."

The light, according to Lewandowski, bounces back and forth from the front surface to the back surface of the glass until it comes to the silicon solar cell, which is embedded within. "Because of this unique design," says Lewandowski, "we could actually get some light down to the back side of the solar cell as well as to the front, theoretically doubling the output of the solar cell...using the same silicon, just the back side as well as the front side." These bi-facial modules are a boon in many respects.

Silicon is the most expensive component in a solar cell, and, Lewandowski says that by reducing the amount of silicon required for energy production by about 40 percent, his company has been able to "replace what's normally $50-a-square-foot material with material that's under $5-a-square-foot." In addition, he says he expects to get to an 85 percent reduction rate in the future, which will drive costs even lower.

Using the current metric, which is based on watts-per-square-meter, the efficiency of the Prism panels is about 14 percent, which according to Lewandowski, is comparable to current PV modules. However, he claims that the Prism panel can "generate over 200 percent more power than a standard silicon solar cell can in cloudy weather or indirect weather." This is due to the fact that the holograms in the glass panels actually act as passive tracking devices, accessing light from various angles as the sun moves through the sky (without any moving or mechanical parts to maintain or repair) and therefore harvesting more kilowatt-hours over time.

Says Lewandowski, "From that perspective, we have a much higher-efficiency product; and, eventually, another year from now we'll have a higher-efficiency module on both levels, using the current [onesun flash test] metric which is based on watts-per-square-meter...plus we'll be able to produce more kilowatt hours." He expects to get up to 18 percent efficiency in the not too distant future: "We've got a lot of interest from companies who are doing third-party financing, so they are setting up farms and they are selling power to the grid. They are particularly interested because we can generate more kilowatt-hours and harvest more kilowatt-hours which means stronger power purchase agreements."

Lewandowski says that, in addition to its grid applications, Prism's current panels "can be a drop-in replacement for any solar module in any application right now; so whether it's commercial, industrial, or residential, you can use them the same way you use the traditional solar photovoltaic module." However, unlike traditional PV modules, Lewandowski says, "The aesthetics [of the Prism Solar modules] are striking. Depending on the angle of the sun and the angle of the module there could be a giant rainbow on the back side of it, and so architects love it. We've gotten calls from very large architects doing projects in Dubai, in China, basically wanting to do a whole building out of this and making a holographic building so that the colors come through."

Building integrated photovoltaics (BIPV), such as the Prism Solar modules, can replace traditional building materials such as roofs, window overhangs, and walls. The National Renewable Energy Laboratory, which has been a leader in solar power development since the 1970s, says that BIPV systems "improve building aesthetics and system reliability while reducing costs and utility transmission losses." One notable example of a BIPV system can currently be found in the Condé Nast Building at 4 Times Square in New York City. In addition to the BIPV solar system, the "Green Giant," as the building is called, also utilizes natural gas powered absorption chillers/heaters, two 200-kW natural-gas fuel cells, energy-efficient lighting, and low e-glass windows.

Lewandowski expects to begin commercial development production of the Prism modules by the end of 2008. In the meantime, Prism Solar is growing exponentially according to its president and CEO. At the beginning of 2008, the company consisted of 24 employees—a figure that Lewandowski estimated would double within a few months. Already, the New York-based company has received funding from the New York State Energy Research and Development Authority, as well as from a California-based venture capital firm.

Its 10,000-square-foot R&D facility in Tucson, Arizona, is working closely with the University of Arizona to continue to refine its product line for the commercial marketplace. In addition to its PV modules, Prism is considering developing a module that makes use of the thermal as well as the PV energy in Prism modules.

"Because we are splitting the spectrum, we [currently] only take the portion that doesn't heat up solar cells," says Lewandowski. "Eventually, we'll build a product that makes hot water and/or heat and photovoltaics because we can split the spectrum in different directions."

Right now, when people think of solar power, they think of California. According to Lewandowski, "California has two great things: it has the market of course, which is backed strongly by the policy-makers out in California, but it also has the [venture capital] community." The other area in the country that has similar market and new venture funding components, he says, is the Northeast: "We see these markets opening up. New Jersey has done great things in developing their marketplace." Other states that have potential untapped markets include New York, Connecticut, and Pennsylvania.

While there is no doubt that he is definitely market-driven, for Lewandowski, solar power is not just about profits: "People need to understand that there's an urgency in this, and it affects not only our economy but our national security, our health. You'd be hard-pressed to look at a portion of your life that renewable energy can't improve."

One of the drawbacks of PV solar is the negative environmental impacts associated with the disposal of solar cells. In 2007, IBM announced that it had developed an innovative reclamation process in which the company can remove and "repurpose" scrap semiconductor wafers. These silicon wafers are the thin discs of silicon material used to imprint patterns that make semiconductor chips for computers, mobile phones, video games, and other consumer electronics. The wafers are also the building block for most solar cells. IBM's process won the "2007 Most Valuable Pollution Prevention Award" from the National Pollution Prevention Roundtable, a non-profit organization devoted to pollution prevention. In addition to the reduction of waste, the process also addresses the silicon shortage, which some say has limited solar panel industry growth.

Other companies are addressing the environmental as well as economic costs of using silicon in solar cells by simply leaving it out of their product line. Konarka is a materials manufacturer based in Lowell, Massachusetts, which makes an organic photovoltaic material (OPV) called Power Plastic that does not use silicon. Konarka has been awarded research contracts from a number of military organizations, including the United States Defense Advanced Research Projects Agency (DARPA).

Konarka research activities include developing lightweight, wearable solar cells, photovoltaic fabrics, printed roofing materials, flexible products, and packaging. In addition to being silicon free, Power Plastic is more sensitive to a broader spectrum of light than conventional solar and can purportedly use all visible light sources—not just sunlight—to generate power.

According to company spokesperson Tracy Wemett, the initial applications for materials will be in three categories: (1) portable power for consumer electronics, (2) indoor applications such as RFID (radio frequency identification) labels and sensors, and (3) outdoor applications such as awnings, tents, and umbrellas. Though not yet as efficient as standard silicon-based solar cells in terms of energy output, the energy required to produce OPV is "negligible compared to what silicon providers need to do," according to Wemett. "There's very little waste."

Currently, says Wemett, "Konarka does have some internal printing capabilities in its own facility to do some modest pilot printing, but the main idea is to scale up and produce with partners." In January 2008, Konarka announced a development agreement with SKYShades, a company that designs, engineers, and installs tension-membrane fabric structures, such as shades and umbrellas, to integrate OPV material into SKYShade products. Konarka will design the Power Plastic prototypes and layout suggestions with an expected delivery sometime in 2008.

On the other side of the continental United States, California's Mojave Desert sizzles with potential energy. The Mojave extends for over 20,000 square miles and takes up a large part of Southern

California. To give some scope to its significance, according to Bright-Source Energy, power plants covering only 1 percent of the Mojave Desert could provide enough solar power for 75 percent of the homes in California and reduce annual carbon emissions by over twenty million tons.

Many companies are looking at the possibilities of solar thermal power, which harnesses the sun's heat to generate electrical power. Solar electricity generating stations are not new to the Mojave. Luz International, one of the pioneers in commercially competitive solar electricity production, designed, developed, built, and operated nine of them between 1984 and 1991, generating a total of 354 MW of electricity. Today, BrightSource Energy and its subsidiary Luz II are working to develop a new type of solar thermal power near Ivanpah, a silver mining ghost town in the Mojave, which lies close to the Nevada border.

While the original Luz International projects were based on solar thermal parabolic trough technology, BrightSource's technology is based on what it calls "dynamic power towers," which the company estimates will increase solar-to-thermal conversion efficiency from about 35 to 40 percent, with lower equipment and project costs.

According to Charlie Ricker, senior vice president of marketing and business development at BrightSource, the 3,500-acre site, which is lcoated forty-five minutes south of Las Vegas, "could go operational in 2010, if we have the transmission access." However, he adds, "It seems likely at this point that we will not have transmission access...and the project will be delayed probably until 2011 for that reason."

BrightSource's dynamic power tower technology consists of a 489-foot tower surrounded by thousands of small mirrors, known as heliostats, which fan out in a solar field. This "solar power cluster" looks much like a high-tech sunflower. The heliostats reflect sunlight to the power tower, at the top of which sits the solar boiler. Says Ricker, "Much as the Boy Scout takes a magnifying glass and makes or starts a fire, it's the same principle that our system works on."

According to Ricker, "Parabolic troughs and most of the other solar thermal technologies use some sort of intermediate fluid. In the case of some of the other companies that use power towers, they typically use molten salt. In the case of parabolic troughs, they use a synthetic oil. We heat water directly to steam. That gives us a significant economic advantage."

A 100-MW power plant will consist of a single tower with a two-stage solar boiler. During the process, the boiler will heat water to 1,100-degree steam in two steps and then pipe it to drive a steam turbine. Although sunlight is plentiful in the Mojave, plants can also be fitted with gas-fired boilers to produce steam at night or on cloudy days. Thermal storage, which would allow heat to be stored during the day and released at night, is also an option, though not a commercially viable one at this point because of the high costs.

The power tower technology itself is well-proven, according to Ricker. In fact, he says the biggest obstacle right now is transmitting the power from the desert into the cities. In the past, power plants were built around the population centers. Says Ricker, "When wind came along, then they started building plants in more remote areas. And then solar is building it up in even more remote areas. So the system, both the physical system and the processes for managing the systems and allocating and so forth were evolved and developed in a day when they didn't face the same sort of issues and problems that they face today. Unfortunately, the processes and the system haven't kept up with the times."

He adds, "In some other states—New Mexico and Texas being probably the most important examples—they have dealt with transmission more effectively. They have gotten out in front of the problem....It takes probably a year or less to get your applications and so forth approved in Texas than it does here. California has not yet gotten out in front of the problems. People want to, but it hasn't happened yet." Ricker believes a permitting system that operates "in the anticipation of need as opposed to post-need" would benefit the state of California, as well as the alternative energy providers who want to do business there.

He is also adamant that property tax abatements and investment tax credits are crucial to continuing the development of more efficient and less costly forms of solar energy. Generally speaking, investment tax credits are designed to encourage economic growth and development by allowing businesses to deduct from their taxes an amount that reflects the amount they have invested in new growth or expansion. A property tax abatement is designed to stimulate new development or redevelopment by temporarily eliminating or reducing the real property taxes a business must pay.

The way Ricker sees it, "If I build an actual gas plant, I would have to buy fuel for the next forty years over the life of that plant, and I would have to buy it every day....If I build a solar plant, I don't have to buy the fuel, but I have to spend a lot more money building the plant. So, in effect, what I have done with that solar field out there is to capitalize the cost of fuel. [I've] replaced an ongoing stream of expenses with a capital cost up front."

Ricker believes this arrangement gives an unfair advantage to fossil fuel producers: "You don't pay property taxes on natural gas. So, if you build this [solar] asset out there and you force it to pay property taxes...then that owner is incurring an expense that you wouldn't incur if you built a plant that used fuel. Providing tax incentives for solar power is a way of leveling the playing field."

While the number of solar technologies and companies that are evolving on the planet today is increasing exponentially, some people are looking toward space as the best venue for developing solar power. The idea of space solar power is not new; however, until recently, it was generally dismissed as being financially infeasible. However, in late 2007, the United States National Security Space Office released an interim report suggesting that, in light of advances in technology and new challenges to security, the United States might want to revisit the opportunities that space-based solar power (SBSP) might offer.

"It appears that technological challenges are closing rapidly and the business case for creating SBSP is improving with each passing year," wrote Joseph D. Rouge, acting director of the National Security Space Office. "Still absent, however, is an appropriate catalyst to stimulate the various interested parties toward actually developing a SBSP capability."

Whether that catalyst will come sooner or later, or even at all, remains to be seen; however, space activists such as Charles Miller, director of the Space Frontier Foundation, a Washington D.C.-based organization dedicated to the exploration and settlement of space, believe the time is now: "The conditions are ripe for something to happen on space solar power," Miller was quoted on CNN.com. "The country that takes the lead on space solar power will be the energy-exporting country for the entire planet for the next few hundred years."

13

Wind Power—A Pioneering Technology

In the 1970s, the idea of providing wind power to cities or communities was a far-out notion shared by a handful of proverbial environmentalists-in-Birkenstocks. Then along came Jim Dehlsen, a "wind power pioneer," according to the United States Department of Energy. Dehlsen had more than a notion; he had a vision. And, with a willingness to take risks and experiment with a nascent technology, he has grown his vision into one of the largest wind turbine manufacturing companies in this country.

In the process, he spurred the growth of the entire wind industry and brought the far-fetched idea of wind power from the fringe to the front of the line in terms of viable forms of renewable energy. To follow his story is to follow the stops, starts, failures, and successes that renewable energy companies—and the people who believe in them—are subject to. With a lot of persistence (and a little bit of luck), Dehlsen, and others like him, are changing the way we look at energy in the United States.

Dehlsen first became interested in energy and the environment during the energy crisis of the 1970s. He formed Zond Systems, Inc., a wind energy development company, in 1980, when wind power was still an nascent technology. The project initially consisted of ten turbines in the southern California desert.

In 2003, Dehlsen recalled, "The performance of these turbines and others we tried was so dismal that we started designing a Zond machine. We also explored the few European turbine options." Dehlsen decided to order 150 Danish Vestas V-15 turbines, each of which produced 65 kW.

Global Warming I$ Good for Business

By 1985, Dehlsen and his crew were working around the clock. However, when the national drive to find renewable energy stalled in the late 1980s, so did United States wind projects, including Dehlsen's. Investment tax credits came to an end, and with them, so too did the prospect of wind as a major part of the energy mix. "Zond went into full survival mode," said Dehlsen. "For the next several years, the only progress Zond made was by reengineering virtually every aspect of the wind projects we had built, with the goal of increasing revenue through greater productivity." Their work paid off: Output increased by over 20 percent.

For Dehlsen, "The central question was how to take wind forward. The answer always came back to driving down the cost of energy." Dehlsen decided to build his own turbines and thus remove the manufacturers' profit margin, as well as to up-scale his machine size. He and his team got "a pivotal break" in 1993 when the Department of Energy's National Renewable Energy Laboratory (NREL) awarded Zond a grant to develop a 550-kW turbine, which was christened in 1995. In an effort to continue to reduce the cost of energy, Zond then developed a variable-speed 750-kW turbine. Said Dehlsen, "Arguably, variable-speed operation was one of the most significant technological innovations since the inception of modern wind energy." Yet still, wind energy was barely a blip in the energy mix.

In 1997, Zond was acquired by Enron, which at that time was one of the largest players in the energy market. People began to take notice. The company acquired Tacke, a German turbine manufacturer, and began to upgrade an even larger 1.5 MW turbine. In 2002, when Enron collapsed, the technology rights and manufacturing assets were sold to General Electric. The newly formed GE Wind Division went on to become the third largest turbine manufacturer in the industry. Dehlsen formed another wind-energy company, Clipper Windpower, in 2002.

The development of incrementally larger and more powerful turbines over two to three decades might seem like a challenge in terms of technology. However, in retrospect, Dehlsen says, "The earliest breakthrough was not in technology as would be expected, but in the creation of a financial method for using the existing state and federal tax incentives for capital formation." Dehlsen also credited the work of government labs such as NREL, which he said, "led and nurtured wind technology toward commercial viability since the 1970s." He added, "In my view [NREL's] work represents one of the best return-on-investments in energy technology ever made by Uncle Sam."

Headquartered in Carpinteria, a small beach community in southern California, with manufacturing facilities in Cedar Rapids, Iowa, Clipper Windpower employs over 600 people worldwide and is engaged in several aspects of wind energy technology, including turbine manufacturing, as well as project development.

One of Clipper's most recent joint-development projects, the Silver Star I, is part of a long-term partnership with BP. The wind project, which is located eighty miles southwest of the Dallas/Fort Worth area, covers about eleven square miles (7,550 acres) of land, across fourteen separate ranches. Because the actual wind turbines will only take up a small percentage of the total area, the land can continue to be used for ranching. The project, which consists of twenty-four turbines, each with a 2.5-MW capacity, provides power for about 20,000 average homes, and most of the homeowners will never know the difference between the wind power and conventional coal power.

In 2007, Clipper received an Outstanding Research and Development Partnership Award from the United States Department of Energy for its 2.5-MW Liberty Wind Turbine, the largest wind turbine manufactured in the United States. The Liberty's unique Quantum Drive Distributed Generation Powertrain is "the next step in the evolution of wind turbine design," according to Steve Lindenberg, acting program manager for the DOE Wind and Hydropower Technologies Program.

In addition to its other projects, in 2007, Clipper announced plans to develop a massive 7.5-MW wind turbine off the coast of England, known as the "Britannia Project." The project, which according to CleanTech Network is the world's largest turbine, "could provide enough electricity for more than 5,500 homes" from a single 7.5-MW turbine.

Says Clipper, "Over recent years, exponential demand for new United States wind energy generating facilities has nearly doubled America's installed wind generation. By the end of 2007, our nation's total wind capacity stood at over 16,000 MW, enough to power more than 4.5 million average American homes each year." At the same time, "a similar surge in wind energy demand has taken place in the European Union countries. Wind power capacity there currently comprises over 50,000 MW." According to the company, "the missing link" for more wind development in the United States "is strong, long-term national policy support—a prerequisite of strong, long-term investment in the sector—that would enable the industry to secure long-lead-time materials and ramp up to employ and train workers to continue wind power's surging growth."

One way in which the federal government has provided incentives for wind and other alternative forms of energy is through production tax credits (PTC), also known as renewable energy production credits. A PTC is a per-kWh tax credit given to "qualified energy resources," according to the Database of State Initiatives for Renewable Energy (DSIRE), an information source on various local, state, utility, and federal renewable energy and energy efficiency incentives. The PTC is essentially a corporate tax credit applicable to the commercial and industrial sectors. Individuals and businesses which produce energy for their own use are not eligible for this credit, though they may be eligible for other tax incentives.

Some critics argue that, while PTCs provide incentives to wholesale energy producers such as large wind farm producers, a feed-in tariff, also known as a feed law, is more equitable for both large and small producers, including distributed energy producers. A feed-in tariff is a mandated premium price paid over a given period (e.g. twenty years) by local electric utilities to renewable energy producers, regardless of whether they are large or small. Proponents argue that feed-in tariffs are more fair and effective than either PTCs or renewable portfolio standards, which require utilities to purchase a given percentage of their electricity portfolio from renewable energy sources. Regardless of which, if any, government incentives are in place, renewable technologies will eventually have to succeed on their own merit. The proliferation of renewable energy companies across the United States is indicative of a real or anticipated demand for alternatives to foreign oil and fossil fuels. Whether or not that demand continues is ultimately up to end-consumers.

While land-based wind farms are gaining more and more acceptance, the Cape Wind project has been a focal point of controversy. However, households on Nantucket Sound will one day see windmills turning lazily out at sea if Jim Gordon, president and founder of Cape Wind Associates, and his team succeed in installing the first offshore wind farm in the United States. But NIMBY (Not In My Back Yard) groups are equally determined to stop him.

After more than seven years, Cape Wind is still a dream for Gordon, but its chances of becoming a reality seem better and better. According to Mark Rodgers, communications director for Cape Wind, the project got the green light from the state of Massachusetts in early 2007 and received a "clean bill of health" from the Department of the Interior roughly a year later. It is moving toward final permitting at the end of 2008, with construction to begin after that.

If everything goes according to schedule, Cape Wind could conceivably be in operation by the end of 2011. Currently, there are several offshore wind farms in Europe; however, offshore wind is still the subject of bitter debate here in the United States, requiring years of environmental evaluation, permitting, and even litigation from opponents who do not want to risk the devaluation of their beach-front property.

According to Rodgers, the company's focus is on obtaining all of the final permits so that building can begin on the 130-turbine wind farm. So far, he says, Cape Wind has the "conditional support" of several leading environmental organizations, including Greenpeace, National Resources Defense Council, Sierra Club and even the Massachusetts Audubon Society. "Many of these environmental organizations, to their credit...are waiting for the end of the environmental process to completely sign off," says Rodgers, who feels optimistic that the project will be deemed environmentally sound and approved for development.

Some of the key areas of concern are the effects of a wind farm facility on bird migration, marine life and the local fishing industry, in addition to opposition from residents for aesthetic and economic reasons. However, says Rodgers, "The most recent poll seems to indicate that there's an emerging majority of support even on the cape and island."

Rodgers estimates that, within the next twenty-five years, the country could get anywhere from 20 to 30 percent of its power from wind, much of which will have to come from offshore. "You've probably heard it said that there is enough wind in the Dakotas to power the entire country," says Rodgers. "The problem is transmission, and it's a really big infrastructure capital cost there." He says transmission is not as big an issue for offshore facilities because they are closer to the major population centers of the United States, especially along the eastern seaboard. In addition, while land-based wind often blows at night, when need is the lowest, offshore wind blows more in the afternoon, when the need is greatest.

This "sea breeze effect," as Rodgers calls it, causes offshore wind to blow more powerfully as the temperature differential between sea and land shifts and air pressure changes. Rodgers estimates that the average output for Cape Wind will be around 180 MW, which is approximately 75 percent of the local power consumed on the cape and island; however, based on data from their weather tower, he says, they averaged 310–320 MW of power during the "top ten electric demand hours," mid-to-late afternoon in July and August, when residents need energy the most.

There is one significant disadvantage that Rodgers points out, and that is the cost of development. Offshore systems currently cost about three times more to construct than their land-based counterparts; however, Rodgers says that may be more than offset by the fact that the cost of wind power is basically stable, once the initial outlay has been made.

"Currently," says Rodgers, "most long-term electricity power purchase contracts in New England are for fossil fuel sources and are for anywhere from a few weeks to about eighteen months at the longest—beyond that, fossil fuel power providers don't want to commit because they do not know their long-term fuel cost....Cape Wind would be able to enter into a twenty-year long-term power purchase contract for either a fixed price or a price indexed to the inflation rate—either would compare well with fossil fuels if people expect their prices to continue to rise anywhere near the rate it has already been rising this decade."

While large energy producers are working to develop onshore and offshore facilities for wind energy, others are focusing on distributed wind energy production. For Southwest Windpower, smaller is better. Their Skystream small wind generator has gained serious venture capital interest and has won such awards as *Popular Science*'s "2006 Best of What's New" and *Time* magazine's "2006 Best Inventions." Redherring.com featured an article entitled "An iPod for Wind Power?" in which the Skystream was likened to Apple's ubiquitous MP3 player in its potential appeal to a large, heretofore untapped, market.

The company says the Skystream 3.7 is the first small wind generator specifically designed for utility-connected residential use. The 21-year-old Southwest Windpower company, "the world's largest producer of small wind generators," now boasts that more than 100,000 of its wind generators have been sold in 120 countries, including the small Air-X wind systems that were sold to 121 elementary and junior high schools in Kyoto.

Closer to home, Southwest Windpower says it has produced small wind turbines for everything from a walkway light at the

Rochester Institute of Technology to off-grid homes on a Navajo reservation in the American Southwest. It also makes the Air-X Marine turbines for sailboats. The secret, according to the company, is to pioneer "new technologies to make renewable energy simple."

However, there may be a catch. Anyone who has dealt with municipal governments can tell you that zoning and permitting takes time, and, because wind energy is still unknown territory for many city planning departments, it can take longer than usual. Diana Hofman of Murrieta, California, reportedly encountered a perplexing situation when she presented a packet of information about Southwest Windpower's Skystream wind generator to the city's planning department. She was told that the department did not have any development standards for wind generators and therefore wouldn't issue a permit without further research and regulations. Hofman was faced with one of two options: either wait "an unknown amount of time" for the city to do the necessary research to establish regulations for permits, or pay up to $4,000 to do the research herself in order to get a permit.

In spite of Hofman's experience, proponents of small wind power claim that regulatory hassles can be overcome with time and patience, and the rewards are more than worth it. Stan and Nancy Arnold of Coal Valley, Illinois, reportedly spent a little less than $10,000 for the turbine, which they bought online. The 33-foot pole was delivered on a semi-trailer, while the 12-foot blades came via Federal Express. Mr. Arnold installed the system himself, following the step-by-step instructions, which included tips for contacting the local utility company and registering with the Federal Energy Regulatory Commission. After all was said and done, the Arnolds cut their energy bill by one-third using their new system, which generated 400 kWh—roughly half of the energy they used—with an average of 10–12 mph winds. And, according to Southwest Windpower, they are not alone. Dottie Neal, of Bovine in West Texas, supposedly saved more than 50 percent on her energy bills in the first year of operation and estimated that the system would pay for itself in less than five years.

One of the main efficiencies of the Southwest Windpower systems is the fact that they require no battery pack or inverter. As the blades turn, the system generates AC current (most systems generate DC current which then has to be converted into AC before powering the home). Furthermore, the turbine can be controlled with a hand-held remote, and, if there is a problem, the remote can be hooked up to a personal computer and data sent to tech support. And, in those areas which have net-metering agreements, wind turbine owners may actually get money back from their utility company.

Aesthetically, Southwest Windpower says its Skystream 3.7 looks much like a common light pole or radio tower. The sound from the turbine "typically blends in with common outside sounds like those from cars, airplanes, barking dogs and wind blowing through the trees." In addition, according to the company, the California Energy Commission has reported that half of the homeowners surveyed would pay more for a home equipped with solar and wind technology.

If Southwest Windpower has its way, windmills may again become a familiar backdrop for the American Southwest and Midwest regions, where winds blow frequently and often consistently enough to make wind energy systems cost-effective. In the words of CEO Frank Greco, "The growing awareness of wind power and its benefits is a clear indication that we are moving toward energy security one household at a time."

While many Americans live in a rural or suburban setting, millions more live in major metropolitan areas. It is rare to see a wind turbine within a large city, mainly due to safety codes and other city regulations. However, according to Bill Jacoby, president of AeroCity, LLC, that may soon change. AeroCity holds an exclusive license to manufacture and sell the Aerotecture turbine, which is uniquely suited to a rooftop environment. The vertical axis turbine, which can be mounted either vertically or horizontally, looks somewhat like an elongated eggbeater within a metal cage. Bill Becker, the turbine's inventor, built eight original turbines, which are still in operation. Since then, he has licensed its use to AeroCity to take the design to commercial development.

"Because we're planning on manufacturing on a much larger scale [than the original eight units]," says Jacoby, "we feel we need to go a little bit more slowly and go through all of the various steps of product optimization and documentation before we start putting them out and see that they get the numbers." Even so, he says demand for his product is so great that it took only three weeks to "line up twenty customers who are anxious to participate as test sites." In addition, he has gotten calls from Massachusetts to as far away as Alaska from people who are interested in installing the turbines.

According to Jacoby, "Most wind turbines would not, in fact, be safe in the city. If you have a fast-turning rotor, you're going to have to worry about ice being thrown or a blade coming off and being flung 500 yards, knifing somebody." However, he adds, "[The Aerotecture turbine], because of its slower rotational speed and the fact that it self-regulates and won't run away at high winds, means at least that it can be installed safely. The second thing is the way it handles variable and turbulent winds....Cities do have turbulent winds, and we have been advised by experts in this that the turbulence at the skyline of a city will actually create, on many roofs, wind conditions which would be quite a bit more favorable [for energy production] than one would guess looking at standard airport wind measurements in a given area."

The birthplace of the original eight Aerotecture turbines is Chicago, also known as the Windy City. Chicago has received a great deal of recognition for its green buildings and technology and even boasts its own green roof on top of its city hall. The rooftop garden is designed to mitigate the urban heat-island effect by replacing a black tar roof with green plants. The results are a cooler roof and less energy needed for air conditioning, as well as a reduced amount of rainwater runoff into the sewers. While not as politically prominent as Chicago's City Hall, the Near North Apartments provide on-site energy in the form of solar panels and wind turbines to residents of the subsidized housing development. Developers reportedly estimate that the wind turbines, which are lined in a row, working together with the solar panels will result in 22 percent less energy consumption than would be found in a traditional building.

The inner helical rotors within the AeroCity wind turbine have some similarities in design to a Savonius Rotor, a vertical axis wind turbine, and are designed to begin producing power at relatively low wind speeds. Says Jacoby, "The power production will continue to increase as wind speed increases, but the RPM levels off." The turbine is housed in an enclosure which stabilizes the rotors, decreasing any tendency to wobble over time and reducing mechanical stresses and noise levels. Jacoby claims that he has stood next to the turbines while they were in operation, and they were "extraordinarily silent."

Currently AeroCity is working with a business acceleration management team from the Hudson Valley Center for Innovation in New York State which, he says, has "ramped up our capabilities." Right now, says Jacoby, "We're continuing to go the early stage venture capital route. We will be seeking [New York State Energy Research and Development Authority] funding and other public grants in the near future."

The AeroCity wind turbines can be used for distributed energy generation either connected to the grid or to a battery bank. Although the turbines currently in use are not building integrated—they are mounted on the roof—Jacoby says that he has heard from architects who want to incorporate the sleek space-age-looking design as tall vertical columns through which the wind can blow and generate power. AeroCity expects to begin manufacturing demonstration models early in 2009. Says Jacoby, "We think once we put one on a rooftop in New York City, it's going to go like gangbusters."

14

Save the Planet: Use More Trees

The best way to stop global warming may be to use more trees, not fewer. So claims Weyerhaeuser, one of the leading producers of lumber and forest products. "For each ton of wood produced by a tree, 1.5 tons of carbon dioxide is removed from the atmosphere," remarked Ernesta Ballard, senior vice president of Weyerhauser corporate affairs, during the 2007 United Nations Climate Change Conference in Bali. "When you make something from a tree, the carbon that is sequestered in forests and the forest products themselves largely offsets the carbon produced by the manufacturing process. Each year, over 100-billion tons of atmospheric carbon dioxide are stored in long-lived wood products. Whether it's basic building lumber or an engineered decorative panel, products from sustainably managed forests hold the answer."

According to Weyerhaeuser, 100 percent of its North American forestlands are certified to Sustainable Forestry Initiative or Canadian Standards Association standards. The Sustainable Forestry Initiative, one of the leading certification programs in the United States, requires program participants such as Weyerhaeuser to undergo an intensive review of its operations by an accredited third-party audit firm in order to ensure the sustainable management of its forest ecosystems as it produces a steady supply of wood and paper products for consumers. Sustainably managed forests not only produce lumber and paper, they also produce by-products that can be used for energy production. Weyerhaeuser says it meets 70 percent of its operational energy needs by burning biomass fuels such as wood residuals and other organic by-products instead of fossil fuels, thus avoiding the release of CO_2 not offset by forest growth.

Global Warming I$ Good for Business

In February 2008, Weyerhaeuser and Chevron took another step toward reducing greenhouse gas emissions by forming Catchlight Energy, LLC, a 50-50 joint venture company, to develop "the next generation" of transportation fuels from renewable, nonfood sources. The initial focus, according to Weyerhaeuser, is on developing technologies to convert cellulose and lignin from different sources into biofuel. One of the more unique aspects of this venture is that cellulosic biomass such as switchgrass can actually be grown on Weyerhaeuser land between the trees that are planted for harvest. Weyerhaeuser spokesman Bruce Amundson explains that, in the Pacific Northwest, new trees may take up to forty to fifty years to grow large enough to be harvested. In the meantime, switchgrass can be planted and harvested every year and can catch enough sunlight between the new tree growth to flourish. Says Amundson, "Historically, our focus in the timberlands area has been solely on growing and then harvesting and then replanting trees that we would use either in solid wood applications like lumber, two-by-fours, panels for construction—that type of thing— or to be made into pulp which was then made into paper, either reading paper or paper used in shipping containers. Now, what we have done…is [found] other revenue streams that we could develop off the timberlands."

Proponents believe that utilizing sustainably produced biomass to make biofuels will answer existing objections to grain-based biofuels. "What our scientists tell us is that, first of all, you can get more biofuel out of wood fiber. You can get a greater quantity at a lower cost than you can from converting, say, corn or other food products into ethanol, and it's much more [energy efficient] from a conversion standpoint," says Amundson. The second advantage is that the process for planting, maintaining, and harvesting trees and switchgrass requires less fossil fuels and emits fewer greenhouse gases than that for growing food crops, such as corn. "This is obviously a nonfood crop and so you don't have…that potential downside," says Amundson. "The other concern people have is that corn today that is maybe shipped to third world countries, or something like that, is now all of a sudden used to fuel someone's SUV in downtown New York." That is not the case with cellulosic ethanol, which is not a grain-based fuel and therefore does not affect grain markets in the same way corn ethanol production might.

Weyerhaeuser's "managed forest plantations" are located primarily in the Southeastern United States, relatively close to Chevron's refineries; however, the process for optimum growth and refining is still in the research and development stage. "It's nothing that's going to help us for next Memorial Day weekend," says Amundson. "It's a number of years out there."

Across the Pacific Ocean, on the sunny island of Hawaii at the National Energy Laboratory of Hawaiian Authority (NELHA), a different kind of nonfood biofuel is being grown—algae. There, Shell Oil and HR BioPetroleum (HRBP) have formed a joint company called Cellana to harness "the power of photosynthesis to address two of the world's major problems—a rapidly diminishing supply of fossil transportation fuels and global warming driven by carbon dioxide emissions." According to HRBP, although there is a great deal of algae-to-fuel research currently in progress, there is little commercial production and none that utilizes the company's proprietary technology. The pilot program at NEHLA is designed to grow marine algae, which yields vegetable oil, for conversion into biofuel.

Proponents claim that algae is one of the most viable forms of biomass available today. Basically, algae needs fresh or salt water, nutrients, carbon dioxide, and sunlight to produce high-quality vegetable oil through photosynthesis. Shell reports that some species of algae may double in size three to four times a day, yielding fifteen times more vegetable oil (per hectare) than rapeseed, palm, or soy. In addition, algae installations can be located virtually anywhere, the main drawback being that commercial-scale operations require large amounts of water.

Algae production is said to be a highly intensive agricultural process. According to Shell, there are basically two methods of producing algae—one is in a closed environment and the other is in open environments where water is pumped through man-made channels or raceways that are open to the air and exposed to sunlight. Both systems rely on water, which must be refreshed each day by pumping it out of the production system and then back in. The advantage of a closed environment is that it is similar to laboratory conditions and takes up less land; however, operating expenses can be prohibitively high. Open algae ponds, like the ones which will be used at the NELHA facility, are less expensive; but they are not as controlled and therefore may be subject to unexpected problems such as water evaporation, temperature fluctuations, contamination, and predators that eat the algae, which only receives sunlight near the surface.

Cellana will use only non-modified marine microalgae species, which thrive in brackish saltwater on relatively small plots of coastal lands, according to HRBP. The algae used will be native to Hawaii or approved by the Hawaii Department of Agriculture. Not only are there no harmful by-products or waste, but algae cultivation could consume over a ton of CO_2 per barrel of biodiesel that is produced. And the by-products of the process can provide supplemental protein for fish or land-based livestock.

Whether or not the same green slime that forms in stagnant pools can be used to get biofuels flowing to the gas pump still remains to be seen, but the creators of Cellana are optimistic. According to Shell executive vice president of future fuels and CO_2, Graeme Sweeney, "Algae have great potential as a sustainable feedstock for production of diesel-type fuels, with a very small CO_2 footprint. This demonstration [at NELHA] will be an important test of the technology and, critically, of commercial viability."

The next time someone asks if you want to buy some swampland in Arizona, you might want to give it some thought. The vast deserts of Arizona have one thing in common with Hawaii—both locales are apparently ideal places to grow algae. GreenFuel Technologies Corporation has partnered with Arizona's largest electric utility company, Arizona Public Service Company, to build a pilot installation at its Redhawk Power Station, located outside of

Phoenix, that recycles industrial CO_2 emissions to grow algae for biofuels. Headquartered in Cambridge, Massachusetts, GreenFuel also has pilot installations in Massachusetts, New York, Kansas, Louisiana, and New Mexico and expects to begin commercial production at an undisclosed location in 2010.

GreenFuel says it uses a "portfolio of technologies to profitably recycle CO_2 from smokestack, fermentation, and geothermal gases via naturally occurring species of algae." A typical GreenFuel enclosed system, which the company calls an "algae-solar bioreactor," resembles a greenhouse. Harvested algae is used to produce a variety of by-products such as dry whole algae, algae oil, and "delipidated meal" that are sold to feed stock and biodiesel fuel producers, as well as others.

Biodiesel is a clean-burning alternative fuel, produced from renewable resources. Unlike diesel fuel, biodiesel contains no petroleum and is made through a chemical process using oilseed crops, such as soybeans. However, the use of soybeans for biodiesel has downsides similar to the use of corn for ethanol in that they both require extensive amounts of land to grow and utilize resources that could otherwise be used to produce food. In comparison, algae biodiesel requires a fraction of the land and resources—some have estimated less than 1 percent of our land could produce enough algae to meet all of our transportation needs—which is one reason why proponents say it is the biodiesel of choice.

While some companies are focusing on commercially viable algae-to-biofuel production, others are giving serious attention to using bugs (as in genetically modified microbes) to produce a sustainable alternative to petroleum-based products. LS9, Inc., based in San Francisco, has endeavored to do something truly unique—design a fuel that looks and acts like petroleum but is, in fact, derived from renewable resources. The renewable resources in this case consist of diverse feedstocks, such as sugar cane or cellulosic biomass, combined with synthetic microbes such as genetically modified E. coli bacteria. The sugars in the feedstock are fermented and broken down into fatty acid intermediates which the microbes then convert directly into "a portfolio of 'drop-in compatible' hydro-carbon-based fuels and chemicals."

The results, says the company, are "a family of fuels that have properties indistinguishable from those of gasoline, diesel, and jet fuel," which essentially means that drivers could conceivably pull up and pump LS9 fuel directly into their SUVs without any engine conversions or modifications. No new pipelines or other infrastructure would be needed since the renewable petroleum is virtually identical to petroleum products in use today, except that it does not produce CO_2 or other smog-related emissions.

In addition to winning numerous awards for its product, the company's global mission (and market potential) have attracted the attention of venture capital firms such as Flagship Ventures—LS9 was the ninth life sciences company started by Flagship—and Khosla Ventures, both of which have been actively involved with the firm since its beginning. The LS9 fuel process is purportedly unique in that it maximizes energy or metabolic efficiency compared with other fuel processes. The fuel produced by LS9's microbes does not need any additional refining and therefore requires approximately 65 percent less energy to produce than ethanol. In addition, because it has a higher energy density than ethanol (its energy density is similar to gasoline), a gallon of renewable petroleum will last about 50 percent longer than a gallon of ethanol. LS9 estimates that its DesignerBiofuels products "will be cost-competitive with $40 to $50 per barrel crude oil without subsidies."

The company has set out to develop its products, first on a test basis and next on a commercial basis. According to Gregory Pal, senior director of corporate development, LS9's pilot facility will be operational in 2008. After that, the company is planning to break ground on a demonstration scale plant in 2009 and then have a commercial scale plant in operation in 2011. Stephen del Cardayre, vice president of R&D at the company, seemed to sum up LS9's strategy when he was recently quoted as saying, "We're dependent on petroleum, so we don't need some alternative to petroleum. We need a way to make petroleum itself."

15

Not Your Father's Automobile

For almost 100 years, General Motors has been offering Americans more than just cars—it has been offering them the freedom of the open road. In a culture that has revolved largely around the automobile, GM has become an icon. However, in the words of president and CEO Rick Wagoner, "Becoming a legend requires innovation. Remaining a legend requires hard work, diligence, and commitment." In a May 2007 speech, Wagoner said, "The global auto industry is beginning to revolutionize the way that autos are powered…and the source of the energy used…and it promises to be the biggest change to hit our industry since, well, the invention of the internal combustion engine."

GM has had to take a hard look at its business model in recent years in the face of increased fuel costs as well as increased competition. It has come up with a strategy for "energy diversity" that includes cars and trucks that can be powered by a diverse range of energy sources. Whether this move toward greener vehicles will give GM the competitive boost it needs to stay in business remains to be seen; however, given the increased demand for fuel efficient automobiles, it may be its best hope. The five categories of "gas-friendly to gas-free" vehicles that GM has proposed include those that are fuel efficient, those that use E85 ethanol, hybrids, electric, and fuel cell-powered cars and trucks. Most of these technologies are still in the testing stages but some are already in the marketplace.

In the near term, the best way to reduce fuel consumption is to make the current internal combustion engines more efficient. GM's "active fuel management" technology allows eight-cylinder vehicles, such as SUVs, to switch to four-cylinder-mode and then back again into eight-cylinder-mode as needed. For example, when the car is idling at a stop light, it might be in four-cylinder-mode and, when it is accelerating up a hill, it could switch into eight-cylinder-mode.

The next near-term technology in the works at GM is the development of flex-fuel vehicles that can run on either gasoline or ethanol. In early 2008, GM announced a partnership with Coskata Inc.—a biology-based renewable energy company—to produce bio-fuels for GM's E85 flex-fuel vehicles. The company estimated that it has about 3.5 million flex-fuel vehicles on the road in the United States, Canada, Europe, and Brazil. Of these, about 2.5 million can operate on up to 85-percent ethanol and 15-percent gasoline (E85) and another million can run on 100-percent ethanol (E100).

In his remarks at the 2008 GM biofuels press conference, Wagoner said that the Coskata process could produce ethanol from any number of renewable source materials, such as agricultural and municipal waste, plastics, and used tires. The process would cost less than $1 per gallon and use less than 1 gallon of water per gallon of ethanol produced. In addition, for every unit of energy that it used in production, the process would create up to 7.7 times that amount of energy. Compared to gasoline, said Wagoner, the Coskata process would reduce greenhouse gas emissions by up to 84 percent. However, one of the main barriers to utilizing the E85 technology is the low number of E85 fueling stations—less than 1 percent of the total number of gas stations in the United States. Nonetheless, Wagoner estimated Coskata would have a pilot plant by the end of 2008 and "a plant capable of producing 50 to 100 million gallons of ethanol a year" by 2011. He claimed, "If all the flex-fuel vehicles that GM, Ford, and Chrysler have on the road right now...plus those that we've already committed to produce over the next twelve years, through 2020...were to run on E85 ethanol, we could displace 29 billion gallons of gasoline annually...or 18 percent of the projected petroleum usage at that time."

Another near-term technology is the hybrid car. When most people think electric hybrid vehicle, they do not think GM. However, GM has been producing two-mode hybrid systems for transit system buses since 2003. In the two-mode system, the first mode is used at low speeds and with light loads, while the second mode is typically used at highway speeds or when pulling heavier loads. Thus, for example, in stop-and-go traffic, a vehicle might run completely under electric power. However, at higher speeds, such as while accelerating to pass traffic, the combustion engine would kick in to provide an added boost. GM believes that its two-mode hybrid technology will pay off in the long run because of its superior performance in larger vehicles, carrying heavier loads or going up hills. The industry apparently agrees. In 2008, the Chevy Tahoe was named Green Car of the Year at the Los Angeles Auto Show.

A longer-term technology involves what GM calls E-Flex electric vehicles, such as the Chevy Volt, which can be plugged into the power grid or run on gasoline, diesel, ethanol, or even (at some point) from a hydrogen fuel cell. These technologies are still in development and, by Wagoner's own estimates, could take "a decade or more" before they have a measurable impact on oil demand. "Everything begins with the battery," said Wagoner at the 2008 Las Vegas Consumer Electronics Show. "The key to getting Volt on the road is advanced lithium-ion battery technology. Our internal tests have shown that individual lithium-ion cells, scaled up to a pack level, will deliver the required power and range....The next step is to begin testing the battery packs in drivable cars." If successful, the Volt will have a regular driving range of forty miles before the gasoline engine kicks in, after which, according to Wagoner, it will "drive several hundred additional miles...with a composite fuel economy of around 150 miles per gallon."

Perhaps the longest-term technology in terms of practical implementation is hydrogen fuel cell technology. Some are skeptical that hydrogen fuel cells will ever be practical. GM disagrees. The company is working on its fourth-generation fuel cell technology and has already begun distributing 100 Chevy Equinox fuel cell electric vehicles to a test market of volunteer drivers in California, New York, and Washington DC. "Project Driveway," as GM calls it, will last up to thirty months and will help GM determine the real-life viability of hydrogen fuel cell vehicles in diverse sections of the country. The EPA certified zero-emission vehicles will be fueled at participating fueling stations, using a compressed hydrogen gas.

At the 2008 Consumer Electronics Show, Wagoner wound up his speech by announcing a fifth-generation fuel cell, using lithium-ion battery technology. The Cadillac Provoq, he said, is capable of driving 300 miles on a single fill of hydrogen. An additional feature is "a solar panel integrated into the roof to help power onboard accessories, such as interior lights and a high-performance audio system."

In many respects, GM has been spurred into action by one of its closest competitors. Toyota Motor Company has a reputation for innovative technologies and "green" manufacturing practices that have set the bar for an entire industry. Bob Carter, Toyota Division group vice president and general manager, recently remarked, "Fuel cells and plug-in hybrids, pure electrics, and lithium-ion batteries and much more will all be part of a future that will require more than just building and selling cars and trucks. It will require a whole new way of doing business." From its corporate commitment to sustainability to its game-changing cars and trucks, Toyota has developed a portfolio of new technologies to address the changing automotive and transportation market.

In many respects, Toyota has been at the frontier of alternative vehicle propulsion technologies for over a decade. The company, which began in the mid-1930s, was founded on "a fundamental belief in balancing business needs with the needs of society," according to Bill Reinert, national manager for Toyota advanced technology. Toyota's "green" approach to both business and the environment has been seen by many as a beacon of light in an industry which has not been known for its eco-friendly products.

In 1997, Toyota introduced a four-door sedan named after the Latin word "to go before." The Prius, as it is more commonly known, has come to change the way American consumers view their automobiles. By 2003, a second-generation Toyota Prius was introduced with an improved hybrid system, known as "hybrid synergy drive."

The Prius hybrid electric vehicle has a gasoline engine which drives both the wheels and a generator, which in turn charges the battery and powers the electric motor. When the vehicle is at a stop sign, the gasoline engine is shut down. When it accelerates at a normal pace, it is powered by the electric motor, which is fed by the battery. When the car cruises at higher speeds (or if the battery wears down), the gasoline engine kicks in. During braking or deceleration, the kinetic energy that is normally lost is instead converted into electrical energy, which is stored in the battery in a process known as "regenerative braking."

In addition, Prius batteries are warrantied to last eight years or 100,000 miles, whichever comes first (in California, they are warrantied to last ten years or 150,000 miles, whichever comes first). When the batteries are ready for replacement, according to a company spokesperson, "Every part of the battery, from the precious metals to the plastics, plates, steel case, and the wiring, are recycled or processed for disposal."

In 2007, Toyota research manager Reinert told the Sustainability Opportunities Summit that the Prius averaged between forty and fifty miles per gallon, twice the mileage of most conventional cars, and produced 70 percent fewer smog-forming emissions. According to Reinert, "We've sold more United States hybrids [in 2006] than Cadillac...Buick...or Mercedes-Benz...sold cars. And more will follow because they are making a big difference. We estimate that all the hybrids sold in America have saved more than 204 million gallons of gas...enough to fill five tanker ships...not to mention keeping more than 4.5 billion pounds of greenhouse gases from spewing in the air."

Although forty to fifty miles per gallon seems like a lot, Oliver Perry says he has seen better. Perry is the president of the Eastern Electric Vehicle Club and organizer of the annual Twenty-First Century Automotive Challenge (formerly known as the Tour del Sol), which "showcases cutting edge approaches to energy-saving automotive transportation." According to Perry, during the 2007 Challenge, one of the competition vehicles—a standard Prius—got 90 miles per gallon using a gas-saving driving technique known as "hypermiling."

By simply driving at more consistent speeds, without the quick stops and starts that so many of us take for granted, drivers can reduce their fuel consumption by 20 percent, according to EPA estimates. Perry says the goal for hypermilers is to get as many miles as they can out of their cars. One way to do that is to "pulse and glide" through traffic at an even speed, accelerating slowly and then coasting to a stop, using the brakes as little as possible. "Everybody could get better gas mileage if they really wanted to get better gas mileage," says Perry. "You have to be conscious of *how* you drive, not just *what* you drive."

While eco-driving can help save gas, more and more people are looking for alternatives to paying high prices at the pump. Toyota now offers several hybrid vehicles, including the V8 Lexus hybrid and the Highlander hybrid SUV, and it is looking at another kind of electric car technology—the plug-in electric vehicle (PHEV). During the 2008 North American International Auto Show in Detroit, the PHEV prototypes that shuttled executives back and forth were designed similarly to the Prius in that they could switch from pure-electric to gas-engine to a blended gas-electric mode. In addition, because they had a second nickel-metal hybrid battery pack, the plug-in Prius vehicles were also able to store greater levels of electricity and thus operate in electric mode longer than the standard hybrid electric Prius, which meant that their tailpipe emissions were also less. At the end of the day, the PHEVs could be plugged into a standard electrical outlet using a three-pronged plug and then charged overnight (the Prius battery is charged by the car engine, not plugged in). Although not currently available to the public, Reinert estimates that PHEVs could be available within the next five to six years.

Yet another alternative fuel vehicle under development is the Toyota Camry Compressed Natural Gas Hybrid vehicle, which operates much like a regular hybrid except that it uses natural gas instead of gasoline. Compressed natural gas, a form of natural gas, is not made from petroleum. It is a clean-burning alternative fuel that is plentiful in the United States; however, getting it from the pipe to the pump is an issue. Currently, there are only about 1,000 refueling stations in the United States, roughly half of which are available to the public. Still, Toyota is optimistic about the chance to open up yet another frontier and says that an expanded compressed natural gas infrastructure could be the model for hydrogen infrastructure further down the road.

By some estimates, fuel cells will take at least ten to twenty years to become a reality. Not only is the technology itself still under development, the hydrogen infrastructure, like that for compressed natural gas, is not yet in place. However, in 2007, fuel cells looked a lot more promising when Toyota engineers loaded up their Highlander Fuel Cell Hybrid Vehicle and hit the Alaska-Canada Highway. One of the problems with fuel cell vehicles in the past has been their inability to function well in extreme cold weather. Another has been their inability to tolerate the vibrations caused by rough road conditions. The engineers chose the Alaska-Canada Highway because it would challenge their vehicle in both of those areas and also because Canada has a network of hydrogen fueling stations that make it possible to refuel every 300 miles. According to Toyota, the road trip was a success. The Highlander performed "without a glitch" for seven days and over the course of 2,300 miles.

While many of Toyota's technologies have focused alternative fuels, the Toyota 1/X (pronounced 1/Xth) compact hybrid concept vehicle is all about size. The 1/X is a plug-in electric vehicle that offers approximately the same interior space as a Toyota Prius at one-third the weight. The vehicle is built from carbon fiber-reinforced plastic, which is lighter and stronger than traditional metals. Also, its roof is made from a bio-plastic material that is derived from kenaf and ramie plants. Toyota claims this car has the potential to travel more than 600 miles on a four-gallon tank of fuel and can be re-charged at home.

One of the things that Toyota looks at very carefully is what it calls the Eco-VAS (vehicle assessment system), otherwise known as the life cycle assessment or well-to-wheels assessment of a vehicle. In 2007, Reinert spoke to a group at University of California, Irvine's CalIT2 Green Information Technology forum. In his remarks, he explained that Toyota does not just look at vehicle emissions—which, in a plug-in hybrid might be zero—it also looks at what materials are used and what is required to gather those materials, to manufacture the vehicle, to fuel the vehicle, to maintain the vehicle, and ultimately to dispose of the vehicle at the end of its useful life.

Under this scenario, a PHEV might have a negative environmental impact if it is powered by electricity from a coal-powered power plant. By the same token, corn ethanol might have a negative environmental impact, given the amount of fresh water and fossil fuels required to plant and harvest the biomass. Said Reinert, "I can make a car with zero emissions, but if I'm creating more emissions to create that fuel, I've just moved the tailpipe back." Without an accurate life-cycle assessment, Reinert explained, "We've got low carbon fuel standards, but we don't have the ability to judge what's a low carbon fuel....Just because a fuel is renewable doesn't mean it's sustainable." Because of this, he said, Toyota engineers are given a "carbon budget" within which they must work, just as they must work within a line budget or a fiscal budget. They may make trade-offs between materials and productions processes as long as they stay within budget.

At the Sustainability Summit, Reinert discussed some of the ways Toyota's Eco-VAS concept extends into the thirteen North American vehicle and parts plants, where environmental performance—including energy consumption, air emissions, and waste disposal—is closely monitored. In addition to reducing energy use per unit of production by 15 percent, all thirteen North American plants have purportedly achieved "zero landfill" status, meaning they reuse or recycle virtually everything. At Toyota's Kentucky plant, workers have used cafeteria food waste as compost for on-site gardens, saving over $1.2 million in waste disposal costs. The Southern California office complex uses recycled carpet, lobby chairs made from old seatbelts, waterless urinals, and recycled water for landscaping. In addition, much of the reinforced steel in the buildings' framework came from recycled cars, and over 50,000 square feet of rooftop solar panels have generated "enough electricity to power 500 homes."

In summing up Toyota's philosophy, Reinert said, "Profitable sustainability does *not* come from random acts of kindness, but rather deliberate actions based on resolve and focus. It takes purpose...discipline...and experimentation...to connect the dots between the needs of earth and the needs of free enterprise."

Although many people are trading in their SUVs and buying hybrids, for some this is simply not an option. Buses and trucks still run on diesel, and that is not likely to change any time soon. Costa Mesa, California-based Extengine, Inc. has received national and international recognition for its Advanced Diesel Emission Control System, "which reduces diesel emissions—including NOx [a major component of smog]—by 80 to 95 percent." The company says that its system can be used on a variety of commercial diesel vehicles, including buses, trucks, and heavy equipment and that it has been verified by California Air Resources Board.

According to a company spokesperson, Extengine's most recent iteration of the ADEC system, which is still undergoing verification by the California Air Resources Board, includes a diesel particulate filter designed to reduce particulate matter—a known trigger for asthma and other respiratory irritations—by 95 percent. Extengine also offers engine lubricants and a biodiesel fuel designed to reduce emissions and improve the performance of heavy vehicles.

What would inspire a former ARCO employee to come up with such a technology? Extengine founder and CEO Phil Roberts was recently quoted as saying, "I guess I got stuck behind too many school buses and trucks."

16

From Railroads to Rapid Transit

For over 100 years, Americans have relied on railroads to travel the huge expanse of land from one coast of the United States to the other. The first American railroads were the modern marvels of their day. Today's rail and rapid transit systems are equally poised to become the modern marvels of tomorrow.

While major automobile manufacturers are making inroads to fuel efficient and even zero-emissions vehicles, other companies are focusing on developing more fuel efficient locomotives. In 2007, General Electric unveiled its hybrid locomotive at Union Station in Los Angeles. Already well-known for its energy-efficient Evolution Series locomotives, GE went one step further and developed a 4,400-hp diesel hybrid electric locomotive engine which includes a bank of lead-free batteries. While not the first hybrid locomotive—Railpower's "Green Goat" wins that distinction—GE's locomotive is the first hybrid "road" locomotive, designed for long-distance travel. Though still in the development stage, GE claims the new hybrid prototype will reduce fuel consumption and emissions by 10 percent, compared to current Evolution Series locomotives.

Hauling freight is big business in the United States, and the Association of American Railroads reports that more than 40 percent of our nation's freight is moved by rail. Today's railroads, like other privately owned companies, are looking for ways to increase efficiencies and reduce costs. According to GE, its new hybrid locomotive will do both. Today's standard locomotives waste a lot of energy in their braking systems. GE estimates the energy dissipated in braking a 207-ton locomotive over a one-year period is enough to power 160 households over the same period of time.

Company engineers have designed a system whereby a series of rechargeable batteries can store that energy when it is not needed and then use it to produce up to 2,000 additional horsepower when it is needed via a computerized control system. GE claims, "If every locomotive in North America could operate as efficiently as GE's hybrid locomotive is being designed to operate, railroads could achieve a fuel-cost savings of $425 million each year."

Hauling freight may be big business, but moving traffic is probably a more immediate problem for the millions of suburban-to-city commuters everywhere. The United States Census Bureau estimates that Americans spend more than 100 hours each year—more than two weeks' worth of vacation time—sitting in traffic. Many would be happy to take the train but balk at the inconvenience of commuting via rail, regardless of how good it might be for the environment. They want faster commute times with less waiting and more convenient schedules, and they are even less likely to consider rail travel for longer, interstate trips because the alternative—airplanes—are so much faster and cost about the same. However, when the cost of gas topped $4 per gallon in 2008, United States commuters began to reconsider their commuting options, and many decided to give local-area rapid transit and rail systems another chance.

In other parts of the world, where cities are closer together (and gas is more expensive), train travel is much more common and convenient. High-speed rail systems are a familiar part of the landscape and magnetic levitation systems are gaining momentum. Conventional and high-speed trains typically use steel wheels on steel tracks, while magnetic levitation (maglev) systems levitate trains off of the track utilizing high-powered magnets.

In France, RailEurope's new SNCF TGV East, a high-speed train, has boasted of speeds of 357 mph, rivaling Central Japan Railway's new maglev test train, which reportedly reached 361 mph. Both of these technologies show just how far the world has come since the Shinkansen "bullet train," the world's first inter-city, high-speed train, traveled from Tokyo to Osaka in 1964 at up to 125 mph. More recently, the world saw another first when the Shanghai Transrapid Maglev Line, the first revenue-producing, point-to-point maglev system, went into commercial service in 2004.

Many people believe that a comprehensive high-speed maglev system within the United States could solve some of our most pressing intra- and inter-state transportation issues. Currently, there are several companies working on maglev technologies in the United States. General Atomics was one of the first. The company's Urban Maglev technology utilizes electrodynamic suspension, in which permanent magnets are arranged in a configuration, called a Halbach array, on the vehicle. When the vehicle moves forward, the Halbach array magnetic fields generate eddy currents on the track that provide a repelling force, levitating the train. The train itself is propelled and guided by a permanent magnet linear synchronous motor, and an electric current is run through the system, creating an electro-magnetic field that smoothly moves the train forward.

Dr. Sam Gurol, director of Maglev Systems at General Atomics, says that typical cruising speeds range from 50 miles per hour for urban passenger transportation to 90 miles per hour for cargo. General Atomics has a maximum planned speed of 100 mph; however, the system could be designed to go much faster—over 300 miles per hour. One of the advantages of an electrodynamic suspension levitation system is its "fail-safe" nature, according to Gurol. Basically, if the power fails, the train will simply glide to a stop before settling back down onto auxiliary wheels. In addition, a train protection system will monitor train speed and positioning using GPS and radar and will automatically shut off power and deploy auxiliary mechanical brakes if those parameters are out of range.

In 2004, General Atomics constructed the first maglev test track in the United States in San Diego, California. The company is now heading a consortium of university, industry, and public utility organization, to develop a demonstration system for passenger service at California University of Pennsylvania (CUP), about forty miles southwest of Pittsburgh. Sponsored by the Federal Transit

Administration, General Atomics engineers hope to start work on the CUP system in 2009 and complete the first phase of operations in 2011. If everything goes according to plan, the system will connect the main campus to the upper campus by 2012, with plans to extend the line out to the city of California, Pennsylvania, after that.

According to Amtrak, rail travel emits about half the greenhouse gases as airplane travel, which makes it a more environmentally friendly alternative to travel by air. Proponents argue that maglev systems are even faster and more energy efficient than conventional diesel locomotives. For example, a conventional passenger train is scheduled to take about eight hours to travel from Boston to Washington D.C. Amtrak's Acela Express, the only high-speed train service currently offered in the United States, takes about six hours to cover the same distance. Compared to that, a high-speed maglev vehicle traveling 100 miles per hour could conceivably cover the same 440 miles in about four hours, only slightly longer than the three to three and a half hours it takes to check in at the airport and fly the same distance.

In addition, the costs to build a maglev system have been favorable, compared with the costs-per-kilometer of building an eight-lane highway in the United States. According to Gurol, a typical eight-lane freeway costs approximately $100 million and can accommodate approximately 12,000 passengers per direction per hour when traffic is moving freely. Maglev systems are competitive in terms of cost-per-mile and can easily accommodate a similar number of passengers; at the same time, maglev systems have a much smaller footprint, with less overall pollution, including noise and tire particulate emissions.

The amount of electricity it takes to operate a maglev train depends on the type of system, how heavy the train is, and how fast it is traveling. According to Gurol, General Atomic's Urban Maglev system consumes about 100 kW of electricity in one hour of operation, which means that a 100-passenger vehicle could run on about $10 of electricity for one hour (assuming a rate of 10 cents per kWh), cruising at up to 50 miles per hour. That electricity can come from any energy source, including off-grid sources such as solar panels along the guideway. Other energy sources might include nuclear, wind, or even a combination of alternative energies. "Of course," says Gurol, "it will be very important as we electrify more of our transportation systems, ranging from cars, trucks/buses, trains, to maglevs, that we choose the power source wisely."

One of the main objections to maglev is the cost of building and then maintaining a working system, which cannot operate on conventional train tracks. Groups such as Orangeline Development Authority, a public-private partnership, not only believe that maglev is feasible, they are actively working to see that this futuristic form of travel becomes a reality sooner rather than later. According to Albert Perdon, executive director of the Orangeline Development Authority, "When you start having higher density development, then the freeways are really not efficient anymore. Land space becomes too valuable, and so now we need to look at transportation technologies that are more in-line with higher density development."

Though still in the planning stages, Perdon believes the Orangeline—which is projected to run 108 miles through the metropolitan Los Angeles area—can not only be accomplished but can, in effect, fund itself. "The way we are approaching it [is] as public-private partnerships or what [California Governor Schwarzenegger] is referring to now as 'performance based infrastructure,' meaning that rather than the government trying to tax everybody and then building a project here and building a project there, that we come up with a project that can fund itself," says Peron. "It's a concept where you build a transportation system that generates enough value for the users that they will pay what it costs to ride on it and cover the operating cost as well as the construction cost."

According to Perdon, the cost of riding the Orangeline will be very competitive to other modes of travel: "I'm not saying everybody is going to give up their automobile and just rely on this project. This is just another alternative, another option. The more people that we can get on systems like this, whether it's light rail or buses or Amtrak or Metrolink...then that takes the pressure off of our freeways. It takes pressure off of the widening of those freeways and the widening of those roads [which] is becoming increasingly expensive. When you have to start buying up very high-priced real estate to add a lane or two to an existing facility, it becomes very costly; so that's why this is becoming more and more a very attractive alternative."

Still, Perdon concedes a project of this magnitude will take many years to get up and running: "The thing that's the most difficult is to build the community consensus, to build the support and get the okays from everybody whose okay we need. That's the challenge....That's the process we are going through. I cannot predict quite honestly if we can develop the necessary consensus from all of the people that are involved, all of the cities, all of the different levels of government, and so forth. I can't predict if we can achieve that in one year or two years or three years." However, he adds, "As long as we keep making progress, then I think it has a good chance of happening."

While some people are working to develop more efficient rail systems between and within our major cities, other entrepreneurs and inventors are endeavoring to redefine the entire concept of transportation altogether. Two such out-of-the-box technologies are dualmode and personal rapid transportation (PRT). While not necessarily new concepts (West Virginia University's PRT has been running for over thirty years), they do present a different—some would say fantastical—solution to the issue of traffic congestion, as well as energy efficiency.

A dualmode transportation system allows for conventional travel on streets and highways as well as automated, computer-operated travel along exclusive guideways. PRT technologies support travel along fully automated elevated guideways with off-line stations for passenger pick-up and departures. PRT's, as their name suggests, typically carry four or fewer passengers, while dualmode carriers can carry larger groups or even cargo.

Kirston Henderson, president of MegaRail Transportation Systems, is a retired systems engineer from Lockheed Martin. He and his team have used their combined years of experience as aerospace engineers to design a "fail-safe" dualmode system which will transport passengers and their vehicles, as well as cargo.

Says Henderson, "You can build all the commuter rail lines...in the world using conventional heavy rail and you would carry 1 or 2 percent of the passengers that are really going to move within an area. The rest of them would all choose to drive their cars because conventional rail systems...are not built on the basis of passenger origin and destination. They are based on where the rail lines go." By comparison, he says, local area MegaRail lines, which would be installed over existent highway rights-of-way, could move passengers in their own cars from stations near their homes to stations near their destinations, "providing them with seamless, same car, door-to-door transport."

Though not a maglev system, Henderson says the MegaRail system can reach speeds of 120 miles per hour, traveling between cities or even across country. He envisions a day when drivers can drive onto an elevated "car ferry," punch in a destination code, then sit back and relax as they are automatically transported on an electric guideway along an existent highway right-of-way above the traffic. When they reach their destination, they simply drive off the ferry and go about their business.

The company is testing a prototype, called a MicroRail system, which is a smaller version designed strictly for travel within cities. Henderson claims it would cost less than 25 percent of what light rail or commuter rail costs. The company also plans to offer a CargoRail technology and has already presented that technology to the Port Authorities of Long Beach and Los Angeles, which are looking for alternative modes of travel to haul freight in and out of the ports.

Chris Perkins, CEO of Unimodal Systems, LLC, comes at rapid transit from a slightly different angle. Perkins was working on a virtual roller-coaster technology at Six Flags Magic Mountain, a theme park in California, when he got the idea of using roller-coaster technologies to improve the way people get around. He pursued the idea and met with Skytran inventor Doug Malewicki. Together, the two began work on their own personal rapid transportation concept.

The Skytran concept is based on high-speed maglev technologies, with projected speeds of up to 150 miles per hour. According to Perkins, Skytran could be deployed virtually anywhere—around town, from city to airport, from city to city or even statewide. Fully-automated pods, each able to carry up to two passengers, would merge on and off line onto the main elevated guideway in much the same way that automobiles merge on and off the freeway via a ramp system. Passengers could board the pods at any one of a number of small, conveniently located stops throughout the city, after which they would enter their destination into an automated system, then sit back and relax as their pod left the station and merged into a network of PRT vehicles.

Skeptics criticize the PRT concept as being technically and economically infeasible; however, Perkins says the company is in the process of determining the location for building a 1,000-foot demonstration loop in California. Eventually, he hopes to build a system of noiseless, lightweight vehicles that can operate 24/7, using "clean electricity," along elevated guidelines throughout the city.

Although MegaRail and SkyTran could be seen as competing technologies, both Henderson and Perkins agree on one thing: They each claim their systems are ready to go and are valid alternatives to the gridlock and energy inefficiencies that even the "greenest" automobiles have failed to resolve.

17

Green Buildings

Until relatively recently, there was no nationally recognized standard for sustainable or green buildings; individual architects, builders, local municipalities, and states designed their own criteria according to their own standards. The United States Green Building Council (USGBC), which was formed in 1993, developed the first LEED (Leadership in Energy and Environmental Design) Green Building Rating System in 2000 to set a national bar by which everyone could measure the projected performance of their projects across a broad range of sustainable criteria. The LEED system was originally designed for new commercial projects such as the Chicago Center for Green Technology; however, it has since expanded to include existing buildings, schools, single family homes, and even neighborhood developments (a pilot program is currently underway involving the USGBC, the Congress for the New Urbanism, and the Natural Resources Defense Council).

The LEED system is a third-party certification program which has become the nationally accepted benchmark for sustainable building design, construction, and operation. A LEED Platinum rating is the highest attainable, followed by LEED Gold and LEED Silver ratings. Each rating is based on guidelines that are set on a consensus basis and are periodically re-evaluated as technological advancements and capabilities evolve in order to keep raising the standard for sustainable development. Since 2000, there have been three updated versions of LEED, and another is coming up for public review in 2009, according to Dr. Malcolm Lewis, former director of the United States Green Building Council from 1997 to 2002. Lewis is currently chairman of the LEED Technical and Scientific Advisory

Committee and a member of the LEED steering committee, which is responsible for the development of standards.

"The thing that's most important is that there is a third-party review," says Lewis. "It's a good check and balance. There's no self-certification." Assuming a LEED consultant is in on the project from the beginning, he says, the certification process can be done in a timely manner, with little additional time needed over and above the length of the project. In addition, he adds, "Many jurisdictions will give accelerated permitting if you're doing LEED certification [as an incentive for builders]."

For new construction, says Lewis, "We're looking at what's in the building—kinds of materials, energy systems, water using systems, etc.—LEED doesn't care what brand you use, but you are asked to submit actual information." Because new construction does not have a proven track record, new construction LEED certification is not based on proven performance as much as on projected performance.

Currently, Lewis estimates about 5 percent of new construction in the United States is being built to LEED standards; however, that percentage does not include LEED commercial interior, core and shell, and existing buildings, which are operations and maintenance projects. "Nobody knows exactly what that number is, but what we've observed is the market has been transformed," says Lewis. "Many of the projects, even if they are not LEED, are still implementing green building practices; so we're really pleased."

Lewis now runs his own technical consulting firm, Constructive Technologies Group, with offices across the United States. Stephen Turner, professional engineer and also chairman of the Technical Committee on Building Environment Designs for the International Standards Organization, is managing director of the Northeast office and has worked with a diverse clientele. "Basically, there are two types of low energy, high performance buildings," says Turner. "There are passive buildings that use low technologies, and there are very sophisticated buildings that use very aggressive technologies. When these buildings use the high tech-approach, there's a lot that can go wrong, and that's our business—to make sure they work well on a systems basis. That's why LEED requires commissioning for any building; it's a prerequisite for LEED certification." He adds, "You could design the most technically advanced building, but, if the systems don't work well together, it may be anything but environmentally friendly."

According to Turner, Aircuity, Inc.'s Center for Green Building Technology is a good example of a technological challenge on several levels: "Many different technologies have been pulled together, so you can not only just see them but you can see them working together. It's unique in that there is a lot of high performance technology for a project of this scale. Even within the high-performance subset of the building sector that Constructive Technologies works in, this is a very exciting project."

Gordon Sharp, CEO of Aircuity, Inc., echoes Turner's assessment. Aircuity is a manufacturer of integrated sensing and control solutions. The systems, which reduce building energy and operating expenses, also improve indoor environmental quality. When the company outgrew its previous facilities, Sharp decided to do more than just move to a bigger space; he decided to move to a greener space. Aircuity leased part of a 125-year-old woolen mill building in Newton, Massachusetts, just outside of Boston, in order to convert the interior of the structure into "a unique, high-profile, highly sustainable facility."

LEED Commercial Interiors (CI) projects are for tenants and others who do not have control over an entire facility but still want to operate as sustainably as possible within their space. Sharp engaged Turner at Constructive Technologies as the commissioning agent for what he saw as a showplace for his own sustainable products and services. Since then, the project has evolved into something more. Says Sharp, "Our goal now is to make it the most sustainable and highest scoring LEED CI Platinum project ever done." To do that, he has decided to include not just Aircuity's Opti-Net systems but other companies' energy efficient and sustainable products and systems as well.

"One of the difficulties you have when you develop a new system or product is people want to see it in a real, live operating environment. You can go to a trade show and see a picture or a mock-up, but most people want to actually touch and feel the real thing and see how it really works," says Sharp. "The thought was to create a facility where we could show off not only our product but also provide it as a forum, a facility for other people to show off their products as well." With over sixty companies now providing their equipment to the facility, the center has become a true showcase for clean technology. Says Sharp, "It's gone from being something for ourselves to something for the industry and the public at large, all of which is very beneficial for everyone involved."

Numerous advanced concepts will be on display throughout the facility, including chilled radiant ceiling panels, under-floor ventilation, hybrid or mixed-mode ventilation, natural ventilation, a solar chimney, automated control of solar shading, solar tracking skylights, rainwater collection for irrigation and greywater, low- or no- water fixtures, an extensive green roof, and, of course, a wide variety of sustainable building materials such as bamboo, reclaimed wood, cotton insulation, and more.

"What's difficult is that we have so many of [these systems]. Any one or two of these would be a design challenge in its own right just to work it out, but having so many of them means there's just a lot of complex issues you're working with," says Sharp. "Effectively, we've taken what would normally fit into five or six buildings and condensed it into one building."

In addition to serving as an office building for Aircuity, Sharp envisions the center as a "living laboratory," where nonprofit and industry groups can perform on-site research using many different energy efficient and sustainable technologies in an operating facility. He also intends to open the facility for tours to industry professionals and the public alike so they can see and experience the comfort conditions of the various indoor environments. Finally, Sharp plans to offer and rent out the facility after hours for "green events," such as meetings, dinners, and receptions, where guests can learn more about green buildings.

In addition to showcasing the latest in clean technology, Sharp intends to showcase some of the building's unique historical features—exposing brick and wood. He is proud that the old mill building was home to Raytheon's headquarters during the late 1940s and early 1950s and that the first commercially produced transistors were manufactured there. "This truly historic facility represents one of the early birthplaces of our modern technology world," says Sharp. "I couldn't think of a more fitting place for a facility dedicated to promoting energy efficient and sustainable technologies that, collectively, have the promise of becoming one of the next major drivers of the world economy in the decades ahead."

Perhaps one of the most revolutionary concepts to come out of the sustainability movement is the idea that human beings can be most productive by cooperating with (versus subduing) their natural environment. Buildings, homes, and even neighborhoods are being planned and built using state-of-the-art sustainable building materials and energy-efficient technologies in order to minimize their impact on the natural ecosystem and maximize the economic well-being (by reducing costly waste) of those who live and work there.

This focus on sustainable systems design has come to include water conservation systems. In the United States, many people take their water for granted, in spite of a growing number of reports that say our natural freshwater resources are dwindling and may soon be unable to meet the needs of a growing population. While some are advocating voluntary and even mandatory conservation programs, others are looking for innovative ways to reduce our water waste and increase the water efficiencies of our homes and businesses.

The award-winning AQUS Water Saver was designed for just that purpose. According to WaterSaver Technologies, "The AQUS is based on the concept that using fresh water to flush the toilet is both unnecessary and wasteful. The AQUS system reduces water waste by reusing greywater from the bathroom sink to flush the toilet, thereby conserving the fresh water that would otherwise be used.

The device was the brainchild of Mark Sanders, a man who hates to brush his teeth with cold water. "My deal is to turn hot water on and let it run while I'm getting my toothbrush and my toothpaste," says Sanders. "I'm watching all this clean drinkable water go right down the drain, and I'm thinking to myself, 'Why can't you somehow take that water and use it for flushing toilets?'"

The system consists of a tank that fits inside a vanity style bathroom sink and can be retrofitted, as well as used in new construction. The tank comes with a basic filtration screen and a tablet dispenser that holds chlorine-bromine tablets to keep bacteria from growing. A twelve-volt DC submersible pump pushes the water up to twenty-five feet to the toilet, where it can be reused to flush.

AQUS, which must pass local inspection, has been used in apartments, single family residences, hotels, office buildings, schools, and state parks and has been sold both nationally and internationally as far as the Middle East. One of the main selling points is that it is small, simple, and relatively inexpensive, unlike larger household systems. And Sanders and cofounder Tom Reynolds are looking into expanding their product line. Says Sanders, "We think the bath and shower tub [version] could be out in the next couple of years."

According to the company, the AQUS can assist builders and developers achieve LEED certification as well as meet their green building goals by reducing wastewater, increasing water efficiency, and contributing to innovation points. And WaterSaver Technologies estimates the AQUS reduces water usage up to 5,000 gallons a year in an average household.

Like Sanders and Reynolds, Jim Poss, founder and president of the BigBelly System, also saw a huge opportunity in reducing waste—in this case garbage. Says Poss, "A couple of facts that jumped out of the page at me were that [collecting garbage] is a $45 billion industry that burns a billion gallons of diesel fuel every year. That's a lot of fuel and that's a lot of money." As Poss calculated the number of trips smog-belching trash trucks had to make back and forth to pick up trash receptacles, he came up with a better idea—he designed a solar powered trash compactor that would handle more garbage and require fewer pickups.

According to the company, the BigBelly is a patented compacting trash receptacle that uses solar power for 100 percent of its energy needs and takes up as much space as an ordinary public trash receptacle; however, its capacity is five times greater than an ordinary trash can. BigBelly cans hold up to 150 gallons of trash and can displace four-out-of-five pickups, which significantly reduces greenhouse gas emissions, air pollution, and fuel costs. And when they are full, a cell phone hook up will download data to a website so that management can arrange for collection.

BigBelly systems can be placed in public parks, corporate campuses, green property developments, and even in shopping malls (they can come with an AC adapter for indoor use). Also, according to the company, the BigBelly system can help earn innovation credits as part of an overall sustainable waste management program for those projects seeking green certification, such as LEED.

Currently, there are several high-profile BigBelly locations including Boston's Faneuil Hall, Walden Pond, and the Alamo. Says Poss, "We are starting to see green property developers and managers...and corporate campuses and things like that. I think it's nice to see the private sector buying into this, becoming more efficient, putting their dollars where their mouths are, doing the right thing. I guess that's what happens when you've got a product with a value proposition that makes sense, not just politically or socially, but in terms of dollars and cents."

Poss, like many "green" entrepreneurs, has discovered that it is not enough just to be eco-friendly. If you can not sustainably solve a problem by providing a quality product that people can afford, you will not succeed. If, however, you can do those things, you may find the results are revolutionary. Perhaps William McDonough, architect, designer, and author, summed it up best: "Commerce," said McDonough, "is the engine of change."

Part Three
The Buck Starts Here

18

Top-Down or Bottom-Up?

Looking back over the past thirty years, it is easy to see the degree to which government regulations, initiatives, incentives, grants, and funding programs have all influenced the way we use our environmental and energy resources. In many ways, the buck *starts* here. The government plays a huge role in stimulating (or squelching) the economic growth in any industry, which in turn affects the financial well-being of us all. At the same time, we also affect the government and the way our representatives may ultimately vote on a given issue.

In the 1960s and early 1970s, there was a great deal of concern about air and water quality, pesticides, toxic wastes, and even the amount of everyday garbage we were dumping into our environment. Citizens waged a large grass-roots campaign, which led to the formation of the United States Environmental Protection Agency, followed by the passage of the Clean Air Act, the Water Pollution Control Act, and other environmental legislation. Then, in 1973, the environmental movement was eclipsed when members of the Organization of Arab Petroleum Exporting Countries placed an embargo on oil to the United States and other Western countries. Supplies dwindled and prices skyrocketed from a national average of 39 cents to a whopping 53 cents per gallon over a one year period. OUT OF GAS signs at gas stations were commonplace, and lines of cars wound around the block waiting to buy fuel on "odd days" or "even days."

In 1977, President Carter declared the energy crisis "the moral equivalent of war" and the new Department of Energy was formed. Congress discussed a slew of bills to reduce our nation's dependence on foreign oil and provided funding for the research and development of nuclear power, coal power, and even renewable energies. Government labs and universities began to explore the costs and benefits of alternatives to gas and oil. Businesses began to spring up to fill new market needs. People began to think about energy in terms of national security. Then, in the 1980s and 1990s, our political woes in the Middle East seemed to improve. Gas prices continued to be high, but supplies were available. We could all drive again whenever and wherever we wanted. There was no apparent need to conserve anymore or to look for alternatives to oil, and most of us went back to our old ways.

Until scientists started talking about something called "global warming."

In 1997, delegates from countries around the world agreed to an amendment to the United Nations Framework Convention on Climate Change, known as the Kyoto Protocol. The primary objective of the protocol was to achieve "stabilization of greenhouse gas concentrations in the atmosphere at a level that would prevent dangerous anthropogenic [human] interference with the climate system." The Kyoto Accord, as it came to be called, required participating developed countries to reduce their greenhouse gas emissions by an average of 5 percent below 1990 levels by the year 2012.

At roughly the same time, the United States Senate passed the Byrd-Hagel Resolution by a vote of 95–0, stating the United States should not sign any protocol that did not bind developing countries such as China and India to the same targets and timetables as developed countries such as the United States. It was the view of the Senate that ratifying such an agreement "would result in serious harm to the economy of the United States." Neither the Clinton Administration nor the Bush Administration submitted the Kyoto Protocol to the United States Senate for ratification. In the words of President Bush, "...the Kyoto Protocol is an unfair and ineffective means of addressing global climate change concerns."

Although Congress did not ratify the Kyoto Protocol, many in Washington D.C. saw the necessity of developing alternative and renewable energy resources and reducing greenhouse gas emissions.

Top-Down or Bottom-Up?

Some began to explore ways to spur a market-driven approach to the development of cleantech and renewable energy sources. In December 2005, former president Bill Clinton reportedly told U.N. delegates that it was "flat wrong" to oppose emissions reductions on economic grounds. He advocated adopting existing cleantech and energy conservation technologies in ways that would enable countries to exceed targeted emissions reductions *and* strengthen their economies. This approach was echoed in the words of General Electric's CEO Jeffrey Immelt: "You can be green and grow."

For most Americans, the concept of global warming did not really take hold until 2006, when former vice president Al Gore presented his movie *An Inconvenient Truth*. The movie, with its easy-to- understand format, outlined some of the issues that many scientists and environmentalists had been discussing for years. It was both controversial and provocative, and it got a lot of people to think about what might be happening to our planet and what we could do to live and work more sustainably. In the end, Gore told his audience: "If we do the right thing, then we're going to create a lot of wealth and we're going to create a lot of jobs because doing the right thing moves us forward."

Although many remained skeptical about the looming prospect of environmental disaster, they did acknowledge that another, closely related crisis was near at hand. In his 2006 State of the Union Address, President Bush admitted, "America is addicted to oil." He went on to announce the Advanced Energy Initiative, which called for a broad range of changes, including a 22 percent increase in clean-energy research and development of alternative and renewable energy sources to power homes and automobiles. The goal, he said, was "to replace more than 75 percent of our oil imports from the Middle East by 2025."

At that time, according to the 2006 Annual Energy Review, petroleum products accounted for 40 percent of the energy consumed in the United States. Coal accounted for another 23 percent, and natural gas accounted for 22 percent. Eight percent of the energy we consumed came from nuclear electric power. And renewable energy came in last, accounting for only 7 percent of the total energy consumed in the United States. However, in a little over two years' time, there are signs that things are changing. The federal government has supported research and development of new technologies both directly through government grants and programs and, indirectly, through tax incentives and public/private partnerships. A *New York Times* article quoted the Environmental Protection Agency as saying, "The United States has invested over $37 billion on climate change science, technology, and tax incentive programs—more than any other country in the world."

The private sector is also upping the ante in terms of developing and deploying clean technologies. The National Venture Capital Association announced that cleantech investments by United States venture capital firms reached $2.6 billion (168 deals) during the first three quarters of 2007, up substantially from the $590 million (63 deals) in 2000. A 2008 Dow Jones VentureSource report says, "The United States is the clear driving force behind the sector's rapid rise, accounting for 83 percent of global clean technology investment in 2007."

In 2007, President Bush announced, "Energy security and climate change are two of the great challenges of our time....We must lead the world to produce fewer greenhouse gas emissions, and we must do it in a way that does not undermine economic growth or prevent nations from delivering greater prosperity for their people. We know this can be done. Last year, America grew our economy while also reducing greenhouse gases." Indeed, according to the Energy Information Administration, total United States greenhouse gas intensity fell by 22.3 percent from 1990 to 2003, at an average annual reduction rate of 1.9 percent.

Emissions intensity is the amount of emissions per dollar of economic activity (gross domestic product). Typically, carbon intensity and greenhouse gas intensity have a similar trajectory because carbon dioxide emissions make up most of the total for United States greenhouse gas emission. The EPA, which keeps a database of carbon intensity per state, claims it is a useful tool that enables states to analyze trends in their carbon intensity and to make comparisons with other states and with the United States as a whole.

With these results, states can effectively target programs and policies to grow their economy and lower their emissions. However, the Natural Resources Defense Council warns that emissions intensity can be misleading if taken out of context. According to the NRDC, emissions intensity can be useful for examining emissions trends, but, in the end, what matters most in terms of the environment is whether or not total emissions have been reduced.

Finding useful tools for measuring progress—let alone deciding what to do once the results are in—is a complex issue, especially in a situation as rapidly shifting as climate change. As with all change, there is a great deal of uncertainty and disagreement about where we will go—or even should go—from here. Many experts believe we should establish national and even global emissions controls and guidelines from the top down in order to effectively deal with climate change; however, others claim that such policies would be ineffective at best and could create a slew of other problems in their wake. For them, the best way to deal with climate change is from the bottom up, with market-driven approaches that will encourage the research, development, and deployment of new technologies—an area in which the United States has historically taken the lead. Even among those who agree that climate change is a real phenomenon to which human beings do contribute, there is disagreement as to what steps we should take to reduce our negative impact on the environment and, perhaps just as importantly, how to measure the effectiveness of those steps in a reasoned versus reactionary way.

In his November 2007 summary to the Senate Commerce, Science and Transportation Committee, John R. Christy, who is professor of atmospheric science at the University of Alabama in Huntsville, director of the Earth System Science Center, and a state climatologist working on Alabama's economic development, questioned the certainty of current climate model projections and advocated a climate science program built on "a foundation of continuous and accurate observations" of climate change as well as a candid appraisal of the methods used to measure and predict change. "We must know *what* the climate is doing before we can understand *why* it does what it does," said Christy. "The topic of human-caused climate change is ubiquitous in the media today. As a result, people are often made to be frightened about the future and their anxiety leads to many state and federal proposals to 'do something' about climate change," However, Christy warned, "It is essential to point out that these scenarios are based on the projections of climate models and are often announced from media personalities whose goals are viewer ratings."

In Christy's view, the government's role should be to continue to support the discovery of new energy sources which address economic and geopolitical as well as environmental issues. Energy security, balance of trade, economic resilience, and air pollution (not to be confused with CO_2 emissions) are all areas which can be explored without the "marginal and uncertain consequences of a desire to 'do something' about climate change." Christy concluded his remarks by saying, "There is no guarantee at all that specific energy policies designed to deal with climate change will actually have the intended effect either in magnitude or sign. Will they produce more or less rain?....No one knows. However, energy policies which address other important issues mentioned above and which include the emphatically desirable goal of affordable energy, and also reduce emissions, are worth pursuing."

In the *Nature* magazine article, "Time to Ditch Kyoto," authors Gwyn Prins at the London School of Economics and Steve Rayner at the University of Oxford also called for a "portfolio of approaches," including an increase in research and development. "Investment in energy research and development should be placed on a wartime footing," they wrote. "It seems reasonable to expect the world's leading economies and emitters to devote as much money to this challenge as they currently spend on military research—in the case of the United States, about $80 billion per year. Such investment would provide a more promising foundation for decarbonization of the global energy system than the current approach."

In addition, Prins and Rayner advocated a "global federalism of climate policy," much like the "United States system of federalism that encourages small-scale policy experiments at the state or local-government levels as well as with the philanthropic and private sectors." When state or local policies succeed, the authors argued, their experiments could more readily be implemented at the federal level "with the enthusiastic support of national politicians." Referring to the December 2007 U.N. Climate Change Conference in Bali, which focused primarily on developing an international consensus for actions to be taken after the Kyoto Protocol expires in 2012, Prins and Rayner concluded, "Although a bottom-up approach may seem painfully slow and sprawling, it may be the only way to build credible institutions that markets endorse. The agenda for the Bali conference should focus on this and on the scale-up of energy research and development rather than on drafting a 'bigger and better' version of Kyoto."

Dr. Michael E. Canes, senior research fellow at the Logistics Management Institute and former vice president and chief economist of the American Petroleum Institute, has compared four basic government policies which can be implemented to control greenhouse gases: Cap and Trade, Carbon Tax, Command and Control (e.g. state Renewable Portfolio Standards and Corporate Average Fuel Economy standards), and Voluntary Goals and Incentives (e.g. government programs and public/private partnerships that set voluntary goals and provide incentives for industry to curb greenhouse gases). Canes advocates strengthening current voluntary goals and incentives by funding public/private partnerships as well as making efforts to deploy technologies to developing countries. He also suggests incentives to adopt energy efficient capital stock such as "more rapid depreciation of new, energy-efficient manufacturing equipment" and "tax incentives to purchase more fuel-efficient passenger vehicles." According to Canes, "The strength of the current United States approach [of voluntary goals and incentives] is that it recognizes that, while curbing greenhouse gases is important, it is only one of many social goals. Growth of gross domestic product is paramount because it provides the wherewithal to achieve all of the goals, including but not limited to greenhouse gas reduction."

James M. Taylor and Joseph L. Bast of the Heartland Institute, a nonprofit organization that promotes free-market solutions, seemed to echo this point of view in 2007 when they wrote: "It is no accident that wealthy countries have made the most progress toward sustainable development. When people are forced to choose between food, clothing, shelter, medicine, or a green environment, a green environment becomes a luxury item. The best way to ensure effective stewardship of the environment is to encourage the development of wealth that makes environmental stewardship possible. Free market [solutions] to environmental protection focus on relying on markets and private property rights to create incentives to protect the environment. Free-market environmentalism is the opposite of the top-down government regulation favored by many environmental advocacy groups."

In short, free-market advocates argue that the United States already *is* taking a leadership role in dealing with climate change by incentivizing market-based solutions and funding the development and deployment of new clean technologies. They argue that it is impossible to mandate innovation. Therefore, we must build solutions to our current environmental problems from the bottom up, by building a robust economy. The more dynamic our economic ecosystem, they claim, the more able businesses will be to develop technologies that will break the relationship between our gross domestic product and our carbon emissions. Only in so doing can we make any significant inroads to reducing our carbon footprint.

Currently, the United States Congress is deliberating climate change legislation that will have long-lasting repercussions. Regardless of whether we decide to mandate national emissions controls or whether we continue to use and augment current voluntary goals and incentives to encourage the growth and development of innovative technologies, it is clear we all have a very large stake in whatever course our government chooses to take next. In the meantime, many in government are already finding innovative ways to encourage the creation of jobs and new technologies that are both sustainable and economically viable. For them, global warming is a challenge that can be met if we show the same determination and willingness to work together that we did when we put a man on the moon.

19

The Moral Equivalent of War

"One of the biggest sources of energy we have is the energy we currently waste," according to Andy Karsner, assistant energy secretary for the DOE's Office of Energy Efficiency and Renewable Energy. For Karsner, "Conservation implies doing less with less; efficiency implies doing more with less." Developing energy-efficient ways of doing things requires more than just conservation; it also requires innovation—namely, coming up with new or improved technologies that will allow us to continue to produce goods and services without any noticeable glitches in supply. This is one of the main thrusts of the Department of Energy.

The agency that was originally begun by President Jimmy Carter as a response to "the moral equivalent of war" now sponsors research and development in several energy areas including clean coal, natural gas and oil, nuclear fission and fusion, bioenergy, geothermal energy, hydrogen energy, hydroelectric or hydropower, wind, solar, and other renewables. Indeed, one of the main focuses of the DOE is "promoting America's energy security through reliable, clean, and affordable energy."

"Many people don't realize that the Department of Energy is the nation's largest sponsor of basic research in the physical sciences—primarily through our network of world-class scientific laboratories," wrote Samuel Bodman, Secretary of Energy. "DOE-sponsored research in both the physical and the life sciences has been a major contributor to the contemporary biotech revolution, which has done so much for our economy and is leading to new scientific breakthroughs all the time."

One of these breakthroughs is bioplastics, which are biobased (versus petroleum based) plastics made from renewable resources. In 2006, Metabolix, which had licensed the basic technology from MIT, and the huge international agricultural processor Archer Daniels Midland formed a joint venture to commercialize the production of a new form of bioplastics called Mirel bioplastics, which are not only sustainable but also totally biodegradable. The company was awarded a $15 million grant from the DOE in 2001 to produce the bioplastics, which are made by microbial fermentation of sugars such as corn or cane sugar or vegetable oils. The end product is resin pellets that can be used to produce everything from personal care products such as disposable razors to gift cards to agricultural films, nettings, and stakes. Although Mirel bioplastics are not yet used for food packaging (food contact applications such as individual packaging need FDA approval), the company does see expanding uses for its sustainable products. It is currently researching the production of bioplastics directly in plants such as switchgrass.

While many companies benefit from government grants, others look to the government for information and expertise in order to attract their own private funding. Within the DOE, the National Renewable Energy Laboratory (NREL) located in Boulder, Colorado, not only develops renewable energy technologies and energy, efficiency practices, but also transfers knowledge and innovations in order to help promote our country's energy and environmental goals. One of the ways it does this is by hosting an annual Industry Growth Forum, a clean energy forum that attracts over 500 participants, including venture capitalists, investment bankers, and cleantech start-ups. The forum features thirty-five start-ups chosen by NREL to present their technologies, which include fuel cells, hydrogen, wind, energy efficiency, photovoltaics, biofuels, and battery storage. Thus far, in addition to gaining an overall insight as to where the clean energy investment market is moving, NREL boasts that presenters during the past five years have raised over $1 billion in financing, due in part to their participation in the forums.

Calling NREL "the NASA for renewables" in his twentieth-annual forum remarks, Paul Dickerson, COO of the Office of Energy Efficiency and Renewable Energy, outlined the long-term policy guidelines that have been set—including the Advanced Energy Initiative's "Twenty in Ten" plan to reduce gasoline usage by 20 percent in ten years—and the research and development that those policy guidelines have inspired. But all of that would be a wasted effort if it weren't for the deployment of technologies, and, according to Dickerson, "You can't talk about deployment without starting at NREL."

In addition to its Industry Growth Forums, NREL also works to transfer innovative technologies to the marketplace through the use of cooperative research and development agreements, licensing agreements, technology partnerships, and even access to its own research facilities and technical staff. NREL's Technology Transfer website provides entrepreneurs with a multitude of resources for bringing their technologies to market as well as an A-to-Z list of over 100 clean energy investors. For more intensive research and development assistance, start-ups can also go to the Clean Energy Alliance, a former DOE/NREL program. The Clean Energy Alliance is a not-for-profit organization that works closely with selected clean energy entrepreneurs in order to help them successfully reach their respective markets. There are over a dozen incubators in states from New York to California, many with ties to universities such as Rutgers, Georgia Tech, and Rensselaer in New York State.

Whether or not these programs and others like them prove effective in increasing energy production *and* reducing greenhouse gas emissions remains to be seen. However, in the words of Paul Dickerson, "It is going to take everyone—the private sector, the government, the scientists, and the consumers all over the globe—to move toward a more sustainable solution."

20

Energy Star

Born out of the grass-roots ecology movement of the 1970s, the Environmental Protection Agency has evolved into one of the most far-reaching organizations to "promote human health and the environment" on the planet. Arguably, one of the most effective ways in which the agency accomplishes its goals is in partnership with the corporations that make the goods and services we all enjoy. According to EPA administrator Stephen Johnson, the agency's voluntary climate change efforts prevented approximately 100-million metric tons of greenhouse gas emissions in 2006 alone—the equivalent of emissions from 60 million vehicles.

Maria Vargas is director of strategic partnerships within the EPA's Climate Protection Partnership Division. She is also the brand manager for Energy Star, one of the best-known programs within the EPA. She takes what she calls a "holistic" approach to crafting a solution to climate change. Since the early 1990s, the Energy Star program has focused on reducing greenhouse gas emissions through energy efficiency. Says Vargas, "The real goal here [is] how can we marry these strides that we have made...in technology? How can we ride that technology train as a way to solve or help reduce greenhouse gas emission?"

Vargas says that the Energy Star program, which includes over 12,000 organizations, is unique because it is a voluntary partnership in which companies not only reduce emissions but also save money on energy. The program is geared toward the consumers who use Energy Star products as well as the retailers who sell them and the manufacturers who make them. "Because Energy Star is a voluntary program, a manufacturer doesn't have to make an Energy Star product," says Vargas. "We set the bar for what it means to be Energy Star and then a manufacturer chooses whether it wants to make a product that earns the Energy Star label." But that label carries a lot of weight. According to Vargas, the EPA has signed agreements with several other countries, including Canada, Japan, Australia, and Taiwan, as well as the EU, all of whom now recognize the Energy Star rating. Because of this, manufacturers only have to put one label on their product to denote energy efficiency, which saves them money.

Energy Star products range from home appliances to office equipment, with more products being evaluated and added each year. The "brand promise" of Energy Star, according to Vargas, is that the model is in the top 25 percent in terms of energy efficiency, that users will not notice any sacrifice in performance—in many cases, they will see better performance—and that it is cost-effective, meaning it will pay for itself in five years or less. The EPA does not issue rebates to consumers who purchase Energy Star-rated products, although many local utilities do. Instead, the EPA has focused more on rewarding long-term changes in behavior by providing consumers with the information they need to buy efficient, cost-saving products.

In addition, the Energy Star program offers a variety of online tools and resources to businesses who want to become more energy efficient. The Energy Star website's "Guidelines for Energy Management" takes businesses through a series of steps, such as assessing performance, creating and implementing an action plan, and recognizing achievements, that will improve energy efficiency. A list of current Energy Star-rated buildings across the United States is also available, as is a list of architects, engineers, and other experts who are familiar with the Energy Star rating system for new buildings. Another useful resource is the "Portfolio Manager," an interactive energy-management tool that allows companies to track and assess energy and water consumption across a portfolio of buildings.

Energy Star offers similar online tools and resources to home-owners. For prospective homeowners, there is a list of Energy Star builders in every state as well as some of the criteria for Energy Star qualified new homes, which include effective insulation, efficient heating and cooling equipment, and efficient lighting and appliances. Existing homeowners can go through an online home energy audit, which assesses how energy efficient their home is and then gives tips for improving that efficiency. Vargas says the biggest challenge they have for existing homes is building "an infrastructure of contractors that can do the work" in a way that will actually reduce greenhouse gas emissions *and* improve energy savings. This requires working with state energy offices, utilities, and others to train contractors and to provide quality assurance and quality control.

Vargas takes her job seriously: "Our acumen is technology. Consumers aren't going to trust something that we can't back up." For her, it's not just a matter of choosing a product, it's a matter of understanding the overall marketplace, the technology options available, and being able to verify and monitor the actual energy savings. "We are data-driven here," says Vargas. "We have to make decisions on what earns the label, and you can only do that if you have enough data to begin to understand where you are going to get the biggest carbon savings, where to invest your time and energy and money." Vargas says her job includes protecting the Energy Star trademark, which consumers rely upon as an unbiased stamp of excellence. That requires keeping a website open, tracking products, and making sure unauthorized companies are not using the Energy Star label.

While the Energy Star program helps businesses and consumers overcome some of the technical barriers to energy efficiency—namely, it helps them identify which products and services are genuinely energy efficient and cost-effective, based on a third-party assessment—the EPA offers other programs to help overcome what Vargas calls organizational barriers to energy efficiency. Sometimes, this requires a fresh look at the organization. For example, is a facility just another overhead cost or could it be a potential investment center?

The EPA's Climate Leaders Partnership program has assisted diverse organizations such as Advanced Micro Devices, Roche Group United States Affiliates, and Xerox Corporation in reducing their greenhouse gas emissions through increased energy efficiency. Each company is unique in the way it operates and each has its own expectations regarding the program. Says Vargas, "Some people need help [setting a baseline]...some people don't understand how best to track their emissions; they want to know what protocols to use."

Vargas and others help organizations set "benchmarks" based on current or baseline annual energy bills. From there, they work together to set achievable goals for future energy and greenhouse gas reductions. At the end of a set time period, they monitor their achievements and look for areas where they can continue to improve.

One of the best ways to reduce energy usage is to make buildings more efficient, an area that dovetails nicely with the Energy Star program, says Vargas. Yet another is to change the type of power you are using within your facilities.

The EPA's Green Power Partnership program encourages organizations to green up their power by buying directly from green power producers, generating their own on-site energy, or purchasing renewable energy certificates (also known as renewable energy credits). Renewable energy certificates are essentially carbon offsets which companies can buy in lieu of purchasing power from a renewable energy producer.

Advocates claim that renewable energy certificates are ideal for companies that are not able to purchase green power from their local provider but still wish to support the development of renewable energy sources. Renewable energy certificates allow those organizations to neutralize or offset their carbon footprint by funding new renewable energy projects, such as wind and solar.

However, critics argue that renewable energy certificates do little to actually support new renewable energy production. A renewable energy certificate, which can be purchased today at $2 per MWh, is simply not enough to justify building a wind farm, for example. In the end, they say, renewable energy certificates do little more than allow companies to advertise themselves as being 100 percent green energy powered, when, in fact they are still using conventional energy sources.

Regardless of the controversy around renewable energy certificates, the EPA says it is possible to make significant progress in reducing actual carbon emissions and increasing energy efficiency. In the words of EPA administrator Stephen Johnson, "America is shifting to a 'green culture,' with consumers embracing energy-efficient products, from cars and televisions to light fixtures and light bulbs." This shift in American culture may be our best hope for sustainability, while we strive to deploy green energy technologies on a meaningful scale.

Stewardship in the Twenty-First Century

Sustainable agriculture brings to mind images of people in nineteenth-century clothing picking organic fruits and vegetables by hand, but today's farmers often use complex machinery and state-of-the-art high-tech to obtain the most consistent yields with the least amount of waste possible. The idea of being sustainable or acting as "stewards" for the next generation is part and parcel of the farming life. Farmers were reducing, reusing, and recycling long before it was popular to do so.

Many old farmhouses—while not technically "green"—are self-sufficient. They are typically about 900 square feet in size (versus the average 2,300-square-foot homes today), with roughly the same size and lay-out as the cutting-edge, energy-efficient homes of the future that are displayed at the Solar Decathlon. These homes are built for function, sometimes consisting of only one bathroom, two bedrooms, a kitchen and family room, and a basement for storage. They are well-insulated against the weather to minimize heating and cooling needs.

Although electricity may come from a local co-op and heat from a propane tank out back, many of the systems on a farm are designed to stand alone. Drinking water is pumped from a well. Sewage is emptied into a septic tank where it is dissolved by bacteria. Water is a precious commodity on a farm. One tub full of water can wash a meal's worth of dishes. Come August, if you have not had much rain or if you have wasted your resources, the well and the cistern will run dry. Then you will either have to do without or load a water tank onto the back of a pickup truck, drive to the water tower in town, and put money into the meter to pump water into the tank, a process which takes hours.

Old farmsteads often have no garbage disposal or dishwasher. Food waste is used as compost to fertilize the garden. Because there is no trash pickup, farmers learn to either reduce their waste (e.g. don't buy individually wrapped items), reuse their waste, or recycle their waste. Taking plastic, glass, and paper to the county recycling centers is a regular weekly trip for many farm wives. The rest of the garbage—and there is very little of it—is usually burned.

When farmers use the term "good stewardship," they are not just being politically correct. Land is more than a commodity—something to be bought and sold and used—it is almost a living, breathing member of their household. They treat it with respect because they know, firsthand, that their own economic survival—and that of their children—depends upon it.

Dr. Jill Auburn is the national program leader for sustainable agriculture within the USDA's Cooperative State Research, Education, and Extension Service (CSREES), which she jokingly refers to as "the agency with the unpronounceable acronym." CSREES is the federal partner for the land-grant university system, a part of the USDA whose functions include research, teaching, and extension. Known among farmers simply as "Co-op Extension Service," it provides training and education for farmers across the country (it also runs the 4H program for youths).

CSREES oversees the federal funds allocated by Congress for land-grant universities as well as for competitive grant programs, and one of Auburn's jobs includes getting that funding to the four Sustainable Agriculture Research and Education (SARE) regions across the country. According to Auburn, the USDA characterizes sustainable development as "sustainable agriculture, sustainable forestry, and sustainable rural community development." Within sustainable agriculture, she says, "Our broad goal is to help agriculture become more profitable and environmentally sound and contribute to quality-of-life for farmers and for communities." Although Auburn admits that there are sometimes trade-offs between the environment, the economy, and the quality-of-life, she says that, "We're aiming for that sweet spot in the middle where you achieve all three of those."

Another goal, says Auburn, is "to help agriculture be a part of the solution to climate change." This is a tall order as more and more people look to agriculture as a source of renewable fuels: "I think one of the big challenges is how are we going to meet society's desire for biofuels and for renewable energy in general in a way that still maintains the soil quality and the other positive attributes of agriculture on the land and maintains food production and the other things that will be competing with fuels." In large part, SARE spurs sustainable agricultural practices through extension education. It is there that researchers, universities, nonprofit organizations, and farmers can work hand-in-hand to implement new ideas and technologies. Not all of the farmers SARE works with are organic farmers, but they are all interested in developing more sustainable and profitable farming practices.

Michele Wander, director of the Agroecology and Sustainable Agriculture Program at the University of Illinois at Urbana-Champaign, is one of the researchers working with the USDA-Integrated Organic Program (IOP). Wander and a team of researchers have been funded to investigate strategies for transition from conventional to organic farming. She also does research that deals directly with tillage conservation techniques and with the use of ground cover or "green manure" to improve soils and crop yields while reducing environmental impacts. According to Wander, the seminal idea behind organic farming and sustainable agriculture is "good soil stewardship." Although the work funded by the IOP deals specifically with organic vegetable farming, much of her research is applicable to conventional row crop farming as well. Ground covers, such as legumes, can be used in both cases and are popular because they "fix" nitrogen in the soil. Wander likens it to "importing free nitrogen fertilizer." Right now, Wander is working "to understand interactions between management and performance because the farmer copes simultaneously with managing the soil and controlling insect pests and weeds while trying to get high-value produce and then trying to market it."

Global Warming I$ Good for Business

Another thing farmers are coping with today is the rapidly increasing cost of fuel and fertilizers. According to Robin Keilbach, a corn, soybean, and wheat farmer in southern Illinois, diesel fuel used to cost $0.19 per gallon in the 1960s and now costs in the neighborhood of $4 per gallon. The costs of petroleum-based fertilizers has also gone up precipitously in the past few years, doubling, and in some cases tripling, in price.

There are several farming techniques that farmers can use to decrease their costs and increase their yields. Minimum-tillage or no-tillage practices, crop rotation, and the use of ground cover (or simply keeping crop residue near the surface where it can turn to humus) are all ways in which farmers can improve the quality of their soil, reduce erosion, and decrease the need for extra fertilizers. Some are also opting to go high-tech and use Variable Rate Technology (VRT). With VRT, soil samples are taken at the rate of one sample of every two-and-a-half acres, each with a given GPS coordinate. Using a computer, the results are programmed onto a computer card, with consideration given to target yield and amount of fertilizer needed for buildup to desired levels. The card is inserted into a computer on the fertilizer application equipment. As the applicator goes through the field (with a GPS uplink), the application rate of the fertilizer varies according to the pre-programmed need and yield. That way, rather than spreading fertilizer across the entire field, the computer will vary the rate and flow of any given fertilizer as needed and skip those areas that don't need it.

Other technologies include planting biotech seeds such as herbicide-tolerant soybeans or genetically modified (GMO) corn that is resistant to certain parasites. This results in fewer pesticides and herbicides treatments to the crops. Says Keilbach, "I get higher yields, lower costs, and it's better for the environment because I'm not putting out as many chemicals." He also claims it reduces the number of trips across the field, which reduces fuel use, CO_2 emissions, and wear-and-tear on equipment. Keilbach says one of the conditions of using the genetically modified corn seed is that he must sign an agreement with the manufacturer to leave a percentage of farmland as a "refuge"—in his case, 80 percent planted in GMO corn and 20 percent planted in non-GMO seed corn—in order to maintain diversity and avoid resistant strains of parasites.

Although the use of herbicides, biotech, and genetically modified crops are a controversial and hotly debated topic, proponents insist that they are necessary to grow enough food to feed an ever-expanding population. Keilbach credits new technology and farming methods with helping him to achieve harvests of 200 bushels an acre of corn today versus 100 bushels an acre in the 1960s. According to the Madison County Farm Bureau, "One Illinois farmer produces enough to feed 146 people in the United States and around the world each year. That number has more than doubled since 1980."

However, organic farmers insist that quality of produce is more important than quantity in terms of human health and sustainability. They claim organic farming nourishes and protects the soil. In addition, by choosing to grow a variety of different crops for market, farmers ensure diversity for their customers and also preserve crop biodiversity.

Every year, SARE selects one farmer from each of its four regions to receive the Madden Award for Sustainable Farming. Some of these farmers, such as Rex Spray of Ohio, used to use conventional farming methods before changing over to organic methods. According to SARE, Spray experienced some initial yield declines, until he perfected his crop rotation, which includes "soybeans, corn, wheat, and hay in alternating grasses and legumes to improve fertility." Now, in addition to the field crops that he and his brother grow on their 680-acre farm, Spray also raises organic beef cattle and has even gone on to sell high-dollar, value-added products such as cleaned and bagged beans for tofu "to maximize profits." At the other end of the farming scale, small farm "specialists" Alex and Betsy Hitt of North Carolina utilize three acres of their five acre farm to grow eighty varieties of twenty-three vegetables, as well as 164 varieties of cut flowers. These high-value crops are in great demand with local chefs as well as at local farmers' markets and co-ops.

Other sustainable farming winners include Maryland dairy farmers Edwin and Marian Fry, who use organic grain to feed their herd and then spread dairy manure over their crop fields for fertilizer. The Frys milk 250 cows and raise 225 replacement heifers each year, and they also sell organic feed to a nearby dairy at premium prices, capturing a niche market in their area. Paul Muller, owner of the 250-acre Full Belly Farm in California, has another approach. He grows a number of diverse crops, such as fruits, nuts, vegetables, and flowers, which he sells through a variety of market outlets. Half of Full Belly's products are sold to retail outlets and Bay Area restaurants, and a quarter sell at local farmers' markets. The remaining quarter goes to Full Belly's 800-member community agriculture enterprise. Community supported agriculture, also known as CSA, is designed to bring local farmers such as Muller together with members of their community. Typically, those in the community who want to buy local produce pledge to support the farming operation. In return for providing working capital for the farm, they receive weekly portions of the harvest during the growing season.

According to Auburn, the main barriers to change for farmers are not only learning new techniques for farming but also taking the time to practice and develop those skills and then actually implement them in the field. For each farmer, the process is a little bit different because no two farms are alike; therefore, farmers often have to innovate ways of adapting new ideas to their particular situation. Says Auburn, "Within agriculture, we see farmers doing incredibly creative things, I mean both to conserve energy but also to produce renewable energy...we see a lot of farmers experimenting with bio-diesel that they can use to power their own farm equipment....There are a lot of farmers who are experimenting with solar and wind power....It's fun to see people taking very diverse kinds of self-sufficiency approaches to their farms and buffering themselves against price rises and the future by really working to make their own operations more diverse and more sustainable."

22

Sustainable Systems

Dr. Tim Lindsey hates waste, not just because of its negative environmental impact but because it is a "symptom of a deeper problem" within an organization. Lindsey is the director of the Pollution Prevention program, part of the Illinois Waste Management and Research Center, which is housed on the University of Illinois campus in Champaign-Urbana. Before working for the state of Illinois, Lindsey worked as an environmental engineer for private industry. It was there that he was introduced to the concept of Total Quality Management. The benefits of increasing the quality and production of goods and services while decreasing waste were obvious. Says Lindsey, "That's where I started connecting the environment to the bottom line."

According to Lindsey, "Pollution is the result of being wasteful, and waste is the result of either a defective process or a defective product....If you look at Mother Nature, it doesn't have any waste. By-products from one organism are food for another and so forth. So what we try to do is get industrial systems to behave like natural ecosystems." Dr. Lindsey and his team work with businesses all over the state of Illinois, helping them re-evaluate their business processes in order to come up with a sustainable solution to their problems: "Once we fully understand the process, then we'll redefine the problem at the root-cause level and then we'll start working on it. And that's something I grill into my people all the time: Understand the process, then understand the problem, then understand the solution. Process, problem, solution. The natural flow."

Lindsey's team has met with some success: "We've worked with hundreds of companies as big as Caterpillar and Ford and as small as one-person operations. They usually come to us." But it wasn't always this way; the Waste Management and Research Center has been around for about twenty years and initially had a difficult time getting companies to engage in its process. Lindsey decided to follow the example of the Cooperative Extension Service at the University of Illinois and actually help companies implement solutions rather than just make recommendations: "I realized after I'd been working about four or five years that we were going around doing these assessments, and people weren't implementing our suggestions even though we could show them that there was no argument about the cost-savings." Lindsey saw that his clients needed more than just another to-do list; they needed "a system for taking the next step."

Over the years, Lindsey says he has learned "if you can get a company to a pilot stage with an innovative technology, practice, method, or whatever, there is about a 70-percent likelihood they'll implement it....If you hand them a list and say, 'By the way, I'm here to roll up my sleeves and help you get this done,' it comes across a lot better." Now, Lindsey says, "It's never been a better time to be green." The number of companies that currently seek his services has noticeably increased. One of the reasons, according to Lindsey, is the research facilities available at the university: "Almost every state has a program [similar to ours], but most of them don't have our laboratories and our ability to do this kind of technical breakdown." Lindsey saw the need for an on-site testing facility where businesses could see if a particular, innovative technology would work before they tried it. Says Lindsey, "It's like you wouldn't buy a car without test-driving it....Once they could test it on site, reduce their uncertainty level associated with it, then they'd feel they could implement it."

Lindsey estimates the Waste Management and Research Center works on about twenty projects a year. One of its relatively recent successes was with a local firm that sold processed vegetable oil "for nominal value" to be turned into animal feed. Lindsey suggested converting the oil to biodiesel instead. "We have extensive laboratory capabilities here, so we developed a recipe for making biodiesel out of their waste grease. And then, once we had the recipe, I worked with their marketing people to take it to the biodiesel manufacturing companies. Not all of them could take it but some could, and they were able to sell that waste grease—and I'm talking about 800,000 gallons—they were able to sell that at probably about three times what they were getting for animal feed."

For Lindsey, who also teaches classes in industrial-mechanical engineering dealing specifically with management and environmental processes, the issue of waste always comes back to sustainable systems. Whether dealing with food-processing waste, chemical waste, or hazardous materials, Lindsey says he always likes to look at the overall process to find out why the waste was generated in the first place: "You started off paying top dollar for these materials, and now [you've] ended up with something that has negative value to you. You have to pay somebody to deal with [it]. Why is that?" In most cases, according to Lindsey, it is a matter of "people not understanding the full cost of being wasteful."

He recalls working with one of the big automakers on Chicago's Southside, which called on him to help reduce its water usage. "Every time I'd make a recommendation to them on how they might reduce water, they'd say 'Yeah, but water is cheap,'" says Lindsey. "So I said, 'Let's back up and really understand what using water costs you in this plant.'…When you looked at the cost of using water as opposed to the cost of just purchasing the water, the costs went from $2.20 per thousand gallons to $80 per thousand gallons.…And, once [the company] understood the true cost of the activities associated with using water, then they really made a lot of changes in the plant. They reduced their water consumption by a third because they understood that the water wasn't cheap anymore. To buy water was cheap, but to use it was expensive."

It is this idea of backing up and looking at the problem from a different perspective that is most intriguing to Lindsey, who says,

"Our next move will be to go beyond processes and into systems because that's really the key to sustainability. If you've got a sustainability problem—like our current energy situation—it's not sustainable because it's a bad system. Retrieving fossil fuels that are hundreds of thousands of years old and dispersing them into the atmosphere is not a sustainable system. So how do you redefine the system and modify the system so that it does make sense? [That's] kind of our next move."

As an example, Lindsey asks, "Do you own a car?" Then, "Do you need to own a car or do you just need to travel? You can ask those types of questions about almost anything." And, according to him, that is exactly what the most successful companies do: "There's a chemical company out east called Haas TCM. They asked the question about fifteen years ago, 'Do our customers need to buy our chemicals or do they just need the performance of the chemicals?' By golly, they decided they just needed the performance; so they stopped selling chemicals and just started to sell the performance....But you can ask those kinds of questions about any product or any service or anything: What does the customer really *need* here? Do they need what I'm selling or [the] performance of what I'm selling?"

Lindsey says one of the worst systems he has seen in terms of waste is the interstate highway system: "You get traffic congestion, so you build more interstate highways, and more people move to the suburbs, and that leads to more traffic congestion. It's a self-perpetuating problem, so that's an example of a bad system." Another example of a bad system is "our current system for supplying transportation fuels." All of this is part of a pervasive cultural misconception. Says Lindsey, "Somewhere along the line, society got the idea that being wasteful was good business." He adds, "What would make you think waste would be profitable?"

The key to sustainability lies in developing a better system. Says Lindsey, "A system can range anywhere from your own home to the world and everything in between, and it's up to you to break it out and see where you can influence the system." For those who have managed to find break-out ways of utilizing sustainable systems, Lindsey claims, "It was more than just recycling their waste or something like that; they actually changed completely the way they interacted with their customers, their suppliers, and everybody."

Still, Lindsey acknowledges that coming up with sustainable solutions is not always easy for the average guy: "We can all make better choices, but we can't make the best choices, and the reason we can't make the best choices is because the systems aren't available to do so....I'm sure [General Motors] will offer fuel cell technology at some point, but they don't now....That's an example of a better system, but I can't participate in that right now....So it's one thing to make choices; it's another thing to have the systems in place that allow the average person to make those choices."

23

Texas: From Big Oil to Big Wind

For years, the state of Texas has been synonymous with images of gushing oil wells and roughnecks wrangling "black gold" or "Texas tea" from the sun-baked earth. Today's Texans, every bit as independent and entrepreneurial as their forebears, are looking at different types of energy to fuel their dreams. Dub Taylor, director of the State Energy Conservation Office (SECO) is one who is working to demonstrate and develop new technologies within the state of Texas.

According to Taylor, SECO steps in after the research and development of new technologies has been done but before those technologies have been deployed in the marketplace: "There is really R&D and D&D [demonstration and deployment], and where we come in is on the D&D side...that's where there may be clean-energy technologies that may have promise, and there may be very good applications somewhere here in the state where we can help find matches."

One recent example is the "cool roof" program. The cool roof technology consists of a reflective coating that can be applied to existing homes and buildings to help cool down the inside of the structure. As Taylor explains it, the idea sounded good on paper; the question was, would it really work? SECO, along with a Texas A & M research team, worked with one of the paint companies to apply the reflective coating to a portable classroom in San Antonio. They measured the heat inside the attic space, compared to an unpainted portable next door. Says Taylor, "It worked as promised and then we did a calculation of the cost-effectiveness, essentially [showing] that there is about a four-year payback—at least in San Antonio—of using a product like this on roof tops." Although perhaps not the most glitzy example of cleantech, Taylor believes the cool roof technology "has a lot of promise for these portable classrooms because they are everywhere....This may be a very cost-effective way to address part of the air-conditioning load for these classrooms that are scattered throughout the state."

SECO has also worked with dairy farmers who are interested in capturing the off-gas from manure to generate electricity. Says Taylor, "We developed an economic model around that to show that you need to have [a certain] number of cattle feeding on this sort of ration to achieve a certain output of waste that you can then convert to energy." With this tool, dairy operators can determine whether or not "they have the opportunity to harvest a waste stream and turn that into something productive."

Another renewable energy project that SECO sponsored was in an unincorporated community of twelve buildings along the United States-Mexico border. Remote communities such as this one are not on any major electrical grid; instead, residents use individual portable gas generators with extension cords for basic lighting and power to their homes. "What we did was package a system that uses a combination of battery storage, a bio-diesel generator, a wind turbine, and solar panels," says Taylor. "So, if the sun is shining, the solar array is charging the battery bank. If the wind is blowing, then that's charging it. And then if those resources aren't available, or if the battery bank drops below a certain level, the generator kicks on and keeps it charged."

Currently, residents are receiving this energy free of charge during testing. However, they will eventually be paying on a pre-paid debit card system, similar to a pre-paid phone card. According to Taylor, "They can take their electricity card, take cash, and buy $50 worth of electricity. They plug in their card to the meter at their house, and the lights are on as long as there is money on the card." Taylor believes this system might work well at remote sites or even in response to natural disasters, where power could be out for long periods of time. "You can drop these units in and essentially have the grid back up [in operation]," says Taylor.

In addition to its demonstration and deployment role, SECO also acts as a "clearing house" for clean and renewable energy as well as energy-efficiency activities. For example, SECO helps coordinate the regional coalitions for the Clean Cities program, which was originally launched in the early 1990s by the DOE. The program was designed to encourage the use of alternative vehicles, as well as the creation of local infrastructures to refuel those vehicles. "The goal...was to support and to create local coalitions. Understanding that this couldn't be managed from the federal level, [the DOE] wanted to enable and mobilize global and regional coalitions to move the ball forward on alternative fuels," says Taylor. "These local coalitions have continued to be pretty active...now [they are] more focused on ethanol, biodiesel, electricity, plug-in hybrid vehicles, things like that. You still have the same need where there are fleets that are making decisions and you are trying to create critical mass of fleets to convince infrastructure fuel providers that the investments they make will be worthwhile." SECO has helped provide funding for fleet conversions and infrastructure installation and has even provided some seed funding for local coalitions.

However, the majority of SECO funding goes to loans for energy-efficiency improvements to public entities, according to Taylor. SECO's LoanSTAR (Save Taxes And Resources) Revolving Loan Program has worked with several cities, including Dallas, which utilized the monies it received to do a "comprehensive retrofit project" on twenty-year-old mechanical systems throughout the downtown area. "We went through and replaced all of those [systems] with newer, more energy-efficient systems," says Taylor. "We optimized the controls and the way that the air conditioning and heating systems were providing heating and cooling to the facilities. And then, in the case of the library, we put in some solar hot water collectors for the domestic hot water needs of the restaurant facilities in the library." The project has a payback period of ten years, and the payback money comes from dollars saved through monthly energy savings.

Taylor, who has a background in the real estate industry, believes "the biggest opportunity for energy savings are existing building stock, whether it's homes or commercial businesses or commercial buildings. So understanding the operating characteristics of those facilities as well as the pressures that the building owners are facing in managing and minimizing their operating costs is probably helpful." By providing up-front funding to those who want to implement improvements but may not have the initial capital outlay, he hopes to spur more energy efficiency and less waste and pollution.

In its efforts to maximize energy efficiency and protect the environment, SECO has developed a network of partnerships, mostly with public entities or local utilities but also including a partnership with the Clean Energy Incubator in Austin. As part of the national Clean Energy Alliance, the Clean Energy Incubator provides the facilities, resources, and expertise to select clean-energy start-up companies in order to help them get funding and ultimately to succeed in the marketplace. A joint program of the Austin Technology Incubator and the IC2 Institute at the University of Texas in Austin, the Clean Energy Incubator has assisted a wide range of start-ups, ranging from small wind to geothermal.

Says Taylor, "I think the higher energy prices here are really starting to spur some activity, not only in energy management and energy efficiency but also in renewable energy as well....You know, we had no wind turbines fifteen years ago, and now we are leading in installed capacity and that trend line is just going straight up." He adds, "Historically, here, we have been an oil and gas state, and now that view of energy is more broad. We are going to continue to be an oil and gas state, but there are a lot of other flavors that we haven't yet fully developed, and those are going to be coming out strong."

Hawaii: The Ocean Provides

Hawaii has long been known for its clear, blue oceans, tropical breezes, and black sand beaches. However, it was not until relatively recently that policy makers and entrepreneurs began to look at those resources as a potential hotbed of energy. In her 2008 State of the State address, Hawaii governor Linda Lingle set the tone for Hawaii when she said, "Our abundant natural sources of energy position us to be a model for the world to show what can be accomplished by developing indigenous renewable energy." Currently, the state is exploring a slew of renewable energies, from wind and solar and geothermal to an energy source that is even more exotic.

One of Hawaii's most unique resources is its ocean environment. Ocean thermal energy conversion (OTEC) utilizes the temperature difference between warm surface seawater and cold deep seawater to produce electricity. OTEC is not a new concept. In the 1970s, it was implemented by the Natural Energy Laboratory of Hawaii Authority (NELHA) at Keahole Point near Kona, Hawaii. The big island of Hawaii is one of the few places where water from 2,000–3,000 feet below sea level can be accessed from land. This cold deep seawater is delivered onshore at 6 degree Celsius (43 degrees Fahrenheit). At the same time, the surface seawater ranges from 24.5 to 27.5 degrees Celsius (76 to 82 degrees Fahrenheit). Experts say the temperature differential between the cold deep seawater and the warm surface seawater could be a great source of potential energy for a state that imports 90 percent of its energy. The DOE's Office of Energy Efficiency and Renewable Energy estimates that "each day, the oceans absorb enough heat from the sun to equal the thermal energy contained in 250 billion barrels of oil. OTEC systems convert this thermal energy into electricity—often while producing desalinated water."

NELHA was originally begun during the energy crisis in 1974 as a support facility for OTEC. The project was scrapped in the mid-1990s; and, since then, NELHA has gone through a series of transformations to become a business incubator for between twenty and thirty businesses. Tenant companies generate products ranging from deep sea mineral water to nutraceuticals to macro algae and aquaculture farms. There are also four water bottling companies in operation and two more waiting for funding, according to Ron Baird, executive director of the facility. In addition to its unique water temperature differentials, NELHA also boasts the highest annual solar isolation of any coastal location in the United States. That, combined with low rainfall and a winterless climate, makes it ideal for growing a variety of organisms, including algae for fuel. As of 2008, Shell Oil's Cellana project is one of two energy projects at NELHA, the other being a solar thermal electricity producer.

Baird is no stranger to business incubators or start-up ventures. He has been in private industry for most of his career, working as an investment banker and private portfolio manager before working for a business incubator that was run by the state of Colorado (he helped to take it private). One of his goals since joining NELHA three years ago has been to develop technologies such as OTEC. Although a new OTEC project was proposed two years ago, it is not yet in operation. Says Baird, "If I had it my way, it'd be in operation now," but he adds, "The approval process and the procurement process has been so torturous that it hasn't yet been done....We want it to be done because it's a technology that can benefit all the people of the world."

Baird's philosophy is simple. According to him, there are three things people need in life: food, healthcare, and energy. Baird believes NELHA is uniquely capable of addressing all three of those needs. "We have companies that are on the leading edge of food production, showing how you can do things beyond anyone's wildest dreams in terms of grazing fish for food," says Baird. In addition, he says, "We have the deepest pipes in the world to be able to go down 2,000 feet. One goes down 3,000 feet. They literally pull up water from maybe 15,000 to 18,000 feet deep." Baird says these waters have been dated by Woods Hole Oceanographic Institute as being approximately 1,250 years old, containing organic compounds that are between 4,000 and 6,000 years old. "There are no man-made impurities in the water," he says. "If pharmaceutical companies think that they were able to go through the rain forests and find all sorts of new things, they haven't even begun to scratch the surface of the ocean."

Baird also thinks the potential for producing renewable energy is great. "I think there are ways we can demonstrate a lot of new technologies and energy that benefit the people of Hawaii," says Baird. "When I say benefit the people of Hawaii, [I mean] anyone who comes in here to build an OTEC plant or do some things in energy that may be very well received in the future or become commercial in the future; we want an equity share in those so that the people get a return on their investment here." Just as Alaskan citizens receive royalties for the energy produced on their land, Baird believes it is only fair that Hawaiians receive a royalty for energy produced on their islands. "It's an alien concept in government," he admits, "but one I think that it's time gets recognized and taken advantage of." Currently, NEHLA does not receive any funding from the state to cover operating expenses, which means the facility has to have revenues equal to expenses. "This has to be operated like a company, a private entity," says Baird. One of the ways to do that is by leasing facilities to private companies.

One of NELHA's newer tenants is the Kona Kai Marine Farm, a division of sixty-year-old Troutlodge, Inc., which is based in Washington State. Jackie Zimmerman, a marine biologist and general manager for Troutlodge marine division, sees sustainable aquaculture as one way to alleviate over-fishing in the world's oceans. "People predict at the rate we are going we will over-fish many of the populations that we have right now if we don't choose to look at an alternative source such as aquaculture," says Zimmerman. "Aquaculture doesn't mean just fish. There's shrimp, there's mollusks, there's bivalves, there's oysters, clams...it encompasses a lot of different seafood products."

Her company is developing a closed life-cycle hatchery in land-based tanks at NELHA. Here, brood stock fish produce eggs which are then hatched and grown from fingerlings into adult fish. These in turn can be used as brood stock for the next generation. Zimmerman is quick to point out that none of the fish are subjected to antibiotics, hormones, or genetic modification to enhance their size or to make them grow faster.

Water temperature and flow has a lot to do with whether or not fish thrive in a given environment, which makes the conditions at NELHA ideal, according to Zimmerman, "because with the cold deep seawater and the warm surface seawater you can combine temperatures in your tanks to whatever you want them to be." Although the farm is evaluating the use of recirculating aquaculture systems for its tanks, it currently uses a flow-through system in which ocean water is pumped in and waste water flows out into a discharge trench beneath the property. There, the water percolates through a lava substrate into the groundwater below before eventually making its way to the ocean. This process, which has been approved by both the county and the state, creates time for the return water to be treated and neutralized in order to eliminate any measurable impact on the ocean, which is an important environmental concern.

Kona Kai Marine Farm bought Aquaculture Unlimited in 2007 and started farming cold-water sablefish, otherwise known as butterfish or black cod, shortly thereafter. Zimmerman says that sablefish are one of the species targeted by the National Oceanographic and Atmospheric Association for aquaculture development. However, Kona Kai picked sablefish to start with because they grow fast for a cold water marine species. Also, the previous owners of the facility had already gone through the time (one-and-a-half to two years) and expense to get permits for two cold-water species, one of which is sablefish. The second permit is for Atlantic halibut, which have a slower growth rate and therefore needs to be tested to see if it is economically feasible to culture and sell.

Two warm water species are also under consideration— amberjack (called kahala in Hawaii) and Pacific threadfin (also known as moi in Hawaii). According to Slow Food International, moi was once a delicacy of the kings and queens of Hawaii and was raised in traditional Hawaiian fishponds known as "loko," along the island coasts. Now, however, "almost all of these ancient fish-ponds are in disrepair" and the moi population "has dwindled to such an extent that a reliable economic supply is no longer avail-able." Currently, the increasing demand for moi is being met by commercial aquaculturalists. And the chefs who offer moi on their menus claim "the quality of the fish raised in these cages is as good as the traditionally raised fish."

One of the most hotly debated issues in aquaculture has to do with the type of feed that the fingerlings eat. Currently, most aquaculture facilities use some mixture of fish oil and fish meal—essentially anchovies or sardines that are caught in the wild—to feed their stock. Some environmental activists claim that aquaculture is anything but sustainable due to the massive amounts of wild fish that must be caught to feed farmed fish. How-ever, according to Zimmerman, aquaculture has strict guidelines just like any other type of agriculture. Although aquaculture is as old as the proverbial seas (by some estimates, it dates back to 2,000 years ago), Zimmerman acknowledges that there are those who will never accept it as sustainable. However, she insists that most aquaculturists are heading in a sustainable direction both because of consumer demand and because of the number of restrictions and regulations they must conform to.

One of the areas in which USDA has shown an interest is in alternate feed sources such as soybeans, wheat, and other non-fish meal choices. "Fish food is a big, big issue when it comes to sustain-ability," says Zimmerman. "It's a big issue when it comes to people being sensitive about what they are eating." Fish food is also the number-one cost for any aquaculture facility. Zimmerman claims, "Substituting fish meal and fish oil [with a grain-based product] would reduce the cost overall of aquaculture [and] make it more economically feasible."

Zimmerman says that aquaculture is growing rapidly throughout the world and believes this is positive as long as it is done in an environmentally balanced way: "To protect our environment, that's what the whole concept was initially in the beginning—to help protect the natural environment...so that we are not over-fishing and removing all of these resources from the ocean without having a way to help it out."

25

Austin: Think Globally, Act Locally

Deep in the heart of Texas lies the city of Austin. Considered by many to be a pioneer in green building, Austin was one of the first cities to recognize and act upon the concept of sustainable growth and development. The headquarters for the Central Texas Clean Cities Program, part of the DOE's Clean Cities Coalition, is located here. Home to the University of Texas, as well as the state capitol building, Austin is a tight-knit community that embodies the spirit of thinking globally and acting locally. Here, citizens and their municipal government are committed to developing clean technologies in both building and transportation that will both spur economic growth and alleviate climate change.

In the early 1990s, when most cities around the United States neither knew nor cared about climate change or energy shortages, Austin was developing a Green Building Program, the first of its kind in the United States. The voluntary program provided basic education, technical assistance, and other resources for customers who wanted to optimize energy efficiency, practice water conservation, utilize sustainable building materials and improve indoor air quality within homes and businesses. By the year 2000, the city council had passed a resolution that required all municipal buildings to meet the United States Green Building Council's LEED Silver rating. Then, in 2007, Austin Mayor Will Wynn announced the Austin Climate Protection Plan, which called for an even more aggressive ramp-up in clean-energy programs that included, among other things, that all city facilities be powered with 100-percent renewable energy by the year 2012. In addition, Mayor Wynn announced another program to make all new single-family homes that are constructed beginning in 2015 "zero-energy capable," or able to produce as much energy as they consume in one year.

At the core of these ideas stands Austin Energy, the municipally-based utility company that serves the greater-Austin area. Ranked as number one in National Renewable Energy Laboratory's 2006 Top Ten Utility Green Power Programs and one of five utilities to earn the DOE's Wind Powering America Public Power Wind Pioneer Awards (2005), Austin Energy is committed to utilizing renewable energies. According to Ed Clark, communications director, 11 percent of the power from Austin Energy will come from either wind or landfill methane gas by the end of 2008, the "vast majority" to come from wind farms in West Texas and the remainder to come from landfill gas capacity. Landfill gas is produced naturally when buried waste undergoes a process called anaerobic digestion. In this case, bacteria break the matter down to produce a biogas. The gas contains about 50 percent methane and can be burned for heat or used to generate electricity, making it a viable form of energy. A side benefit to harvesting landfill gas is that the methane that would normally spill out into the atmosphere is used to produce energy instead.

However, the real "source" of energy in Austin does not come from a power facility but rather from the cumulative energy savings from the Greenbuilding and Energy Efficiency programs that have been in place since the early 1980s. In his testimony before the House Committee on Small Business, Wynn said, "In 1983, Austin Energy...was on the verge of building a new coal-fired power plant on the outskirts of the city. Our citizenry voiced strong opposition and city leaders responded by launching an aggressive energy efficiency and conservation program. To date, that program has eliminated the need for more than 600 megawatts of electric generation capacity. And that coal-fired plant was never built." There are two aspects of the program which work together but focus on different groups of people. Green building focuses on building professionals such as architects, engineers, and contractors in order to get energy and water efficiency measures built into new homes and businesses. Energy efficiency is geared for people to make their existing homes and businesses more energy efficient. Clark notes that both green building and energy efficiency are "a major component of the way we address new load growth....It's not just something we do to feel good." The overall concept, he says, is a real win for consumers as well as for the environment because "if you don't build a plant, your utility bills will stay lower."

Clark estimates about 20 percent of the new homes built in Austin each year are "green rated" for efficiency. "Green building is maximizing every component of the way a home [or a building] is built. The way it faces, what kind of insulation is used, what you do to your roof to reflect sunlight back out so that that doesn't become a heat source for the home or the building...what kind of equipment you put in if you are heating or cooling." Chemical-free and recycled construction materials are also a consideration, as is on-site waste and energy management. Austin Energy offers one of the largest if not the largest rebate program for solar installation in the country, says Clark. And the Austin Water Utility offers 75-gallon barrels, debris screens, outflow hoses, and overflow tubes for customers who want to use rainwater for their gardens and lawns.

Austin Energy gets no tax money; its revenue comes from the sale of electricity. "Electric utilities are in the business of selling power, and we are too; so it is a little bit of a change to put this much of an emphasis on using energy efficiency," Clark admits. But he believes the cost savings from not having to construct new power plants makes a big difference that can be passed on to consumers through a series of rebate and incentives programs. It can also be put into useful hands-on programs, such as opening up the grid as a test bed for Austin's Clean Energy Incubator as well as for some of the cleantech start-ups that share its facilities.

In a 2007 hearing before the House Committee on Small Business, Austin's Mayor Wynn claimed, "We have the fastest growing economy of any city our size. Moody's recently rated Austin the number-one city in the nation for economic vitality. *Forbes* consistently rates us among the top cities for business and careers. *Harvard Business Review* rated us the number-one city for business creativity last year. The list goes on. We're living proof that forward-thinking energy policies and strong environmental protections aren't in competition with economic strength but, rather, are complementary to it."

26

Portland: From Stumptown to Sustainable

According to local lore, Portland, Oregon, was given the nickname "Stumptown" during a period of rapid growth when untold numbers of tree stumps were literally left where they had been cut in the rush to make way for a new city. More recently, Sustain-Lane, a web-based guide to sustainable living, ranked Portland number one in sustainability and called the city, "A Role Model for the Nation" for its city innovation, green economy, and overall knowledge base that "reflect a deep-seated understanding of sustainable practice." The journey from Stumptown to sustainable is an interesting one. At its core, is a commitment to protecting the natural environment in an economically rewarding way.

In his January 2008 State of the City address, Mayor Potter said, "We have reduced greenhouse gas emissions by 14 percent per person and at the same time increased jobs by 14 percent—proving a clean environment can create jobs and make money for hundreds of local companies." In addition to being leader in LEED certified buildings, Potter also claimed that Portland has the highest combined residential/commercial recycling rates (63 percent). But the city's commitment to green energy does not just stop at its borders. "Portland service stations now pump biodiesel. And Portland's actions are creating jobs not just here, but in other parts of Oregon," said Potter. "Our biodiesel is being produced from canola from eastern Oregon and potato chip oil from central Oregon. The city's wind power that will fuel 100 percent of city government needs by 2010 comes from the windy hills of eastern Oregon."

Global Warming I$ Good for Business

Susan Anderson, director of Sustainable Development for the city of Portland, is proud of what her city has accomplished. Portland first began tracking its CO_2 emissions in the early 1990s. As one of the first cities in the United States to do that, they "took some great pains to figure out a methodology for how to do it." Basically, according to Anderson, they added up all the fuel used by the city, including natural gas and electrical (kWh) usage amounts, as well as transportation fuels used, and then converted those numbers into CO_2 emissions figures. As members of the International Council for Local Environmental Initiatives, they can now compare their emissions with those of other council members as well as exchange ideas on how to track and reduce fossil fuel use.

According to Anderson, "The key most daunting issue I think for the United States is land use and how do you create an urban environment [where] you can use a whole lot less resources but still have the same quality-of-life or even a better quality-of-life." One of the ways that Portland has endeavored to do this is by reducing waste and by recycling. While the average city in the United States might recycle only around 30 percent of its waste, Portland recycles twice that amount. As of May 2008, Portland has new mandatory programs in place whereby food waste from restaurants and other institutions must be recycled for composting. Similarly, wood waste and debris from construction and demolition must be reused or recycled. Also, the city has worked extensively to establish a solid base of architects, engineers, and other professionals who are trained to build LEED-certified buildings. With "over 150 LEED-certified buildings either underway or completed in the metro area," Anderson says that building a Gold LEED-certified facility "is not a big deal here anymore...the market has moved really quickly in the Northwest."

The city is also considering a carbon "fee-bate" to reward those who build new construction residential and commercial buildings above the energy code. "We are going to propose that if you just meet the energy code when you do any construction, you basically will have to pay a fee based on the amount of carbon dioxide emissions...related to that structure over a certain amount of years," says Anderson. "If you go 30 percent beyond the code, the fee will be waived, and if you go 40 percent beyond the code, then we will actually pay you." Anderson says, "There are definitely builders already building to the higher level...so it's very doable." For Anderson, one of the keys to sustainability is to work in partnership with local businesses. "It's definitely not a government-run kind of project," she says. "There's a mix [of] government, nonprofit, and business neighborhood organizations coming together."

In order to continue to encourage sustainability, Portland has also begun a Green Investment Fund (GIF) for "cutting-edge projects." In 2007, Commissioner Dan Saltzman, who first introduced the GIF concept, remarked, "The Green Investment Fund is a prime example of what we can actively do to promote innovation in the field." Projects such as a community center constructed from shipping containers and a multifamily residence with a "living wall" that processes storm water are two of those projects which have received up to $425,000 in funding. The Energy Trust of Oregon, an independent nonprofit organization, is one of the fund partners and is actively involved in promoting energy efficiency and renewable energy throughout the state. The state of Oregon, while not directly involved with GIF, is also a factor in green development, according to Anderson. Energy tax credits, which allow businesses to take a predetermined dollar amount per-square-foot deduction from their income taxes, provide a strong motivation to build sustainably.

One of the areas of focus for sustainable growth and development is water drainage. In a city where rain falls over 150 days per year on average, problems with overflow are a major concern. "The whole city, just like every urban area, has been paved almost everywhere," says Anderson. "So now we are looking for simple changes and cutting-edge solutions for making everything more permeable." In addition to building a $1.2 billion pipe system to keep water from overflowing the Willamette River or the city sewage facilities, Portland is helping to building eco-roofs [green roof systems] and swales [grassed channel drainage ways] that will slow the flow and allow the city to "get the value out of it."

Although Portland has made a point of growing sustainably, Anderson sees more room for improvement. "The issue is huge, and to get there it's not going to be screwing in a few lightbulbs and it's not going to be increasing car miles per gallon [by] 5 percent or something. It's basically a transformative change that happens, that needs to happen, to the way urban environments work," she says. "It's daunting and yet possible, or maybe I'm just an optimist." In a city where the visitor's association website features green vacation deals such as the "Go Green" and "Carless Vacation" packages, many people seem to share her optimism.

27

Miami: Smart Growth

Smart Growth is at the core of Miami's strategy for climate change, not only in terms of green buildings but also in the way the city is actually laid out. Miami 21, which stands for "Miami of the twenty-first century," is at the cornerstone of the city's plan for smart growth. In his 2007 remarks before the House Select Committee on Energy Independence and Global Warming, Mayor Manuel Diaz said, "Miami 21 is rooted in the belief of the power of traditional neighborhoods to restore the functions of sustainable cities. It strives to achieve a unique sense of community and place, challenging old assumptions in urban planning by providing an alternative to urban sprawl, traffic congestion, disconnected neighborhoods and urban decay."

Basically, the Miami 21 concept calls for a zoning overhaul— from land-use zoning codes to form-based zoning codes—which "entails a holistic approach to land use and urban planning." Mixed-use areas blend retail, office, and residential uses within pedestrian-friendly buildings with landscaped walkways, plenty of windows along the street, and internalized parking and loading space to encourage pedestrian activity. Additionally, the form-based code creates nodes of building intensity that promote transit usage, redevelopment, and reduce the need for cars. From that, city planners hope to create a sense of neighborhood which has all but disappeared in most modern-day communities. In addition, they expect to reduce the number of commuters from homes in the suburbs traveling to downtown offices.

According to Alex Adams, urban design planner and project manager for Miami 21, the core of this philosophy is the concept of transect zones. Transect zones relate the environmental and urban components of regional and city planning. Instead of defining an area by the uses of buildings, planners can instead look at the inclusive functions of a zone in relationship to the environment. Transect zones occur as varying degrees or gradients of an urban-to-environment ratio beginning with "(T1) Nature," such as the Everglades, transitioning to "(T3) Suburban," a mix of nature and buildings, to "(T6) Urban," centers, where the urban form is dominant.

"Development," says Adams, "should be built as nodes, so building intensity increases around those nodes, which ideally will be around transit routes." As an incentive, he says, builders are given a base level of buildable rights and then allowed a bonus as an incentive for building sustainably. There are five qualifying bonus criteria that will earn a bonus for the builders: (1) affordable housing, (2) historic preservation, (3) brownfields, or building in areas that have fallen into disuse or disrepair, (4) green building (under Miami 21, buildings over 50,000 square feet are expected to be build to a minimum LEED Silver rating), and (5) parks and open spaces.

With the bonus program, for example, builders who are allowed 50,000 square feet of buildable rights, may earn the right to build an extra 10,000 square feet if, say, they build to a LEED Gold standard instead of the minimum LEED Silver. By the same token, if builders endeavor to preserve an historic building, they many earn bonus points which can be transferred to a new construction project. For example, generally speaking, neighborhood nodes are allowed a smaller bonus (up to 25 percent) compared to downtown urban nodes, which are allowed a larger bonus (up to 50 percent) because, Adams explains, "we want to encourage building in the city center." Another way builders can earn buildable rights is to voluntarily contribute money or land to a city trust fund, which goes into affordable housing, brownfields redevelopment, or the establishment of parks and open spaces. "Basically, we don't care why builders build sustainably," says Adams. "We just want to make this a better city."

Miami: Smart Growth

Although transportation is a county and state issue, according to Adams, the city is doing what it can to encourage less reliance on automotive traffic and more on alternative modes of transportation. In 2008, Adams says, a Bicycle Action Plan was put into effect to help develop bike friendly corridors that link employment centers, parks, schools, and other facilities. Mayor Diaz affirmed this commitment to alternative transportation. "For far too long, cities have been planned around cars and not people. Government policies have invested in sprawl by encouraging the use of cars," he said. "Miami 21 brings sustainability through design. It re-imagines Miami in a way that makes sense to pedestrians, so our city is no longer subservient to cars. It will also offer transportation alternatives, including a return to streetcars like the ones Miami had several decades ago." In fact, Miami was one of the first United States cities to build an automated downtown people-mover in the 1970s, and the city still operates the Miami Metromover—an elevated, fully-automated system—within the downtown Miami area. However, until relatively recently, the city has been recognized more for its weather than for its sustainability.

As with so many other southeastern United States cities, Miami is experiencing one of the driest times in its history. The City of Miami estimates the average person in Miami-Dade uses 158 gallons of water a day, almost half of that outside the home to maintain landscapes. Water conservation is a key issue in this city, as is the use of native plants which are also drought-tolerant. "In a city where hurricanes and other forces have depleted our tree cover, canopy replacement is an issue of concern," remarked Mayor Diaz. "We have a city arborist and are planting native species that provide ground cover, convert noxious gases to oxygen, and lower ground temperatures." The city has also embarked on an outreach program called "One Person, Ten Steps, Ten Tons," which gives ten easy steps that everyone can do to reduce carbon emissions by 10 tons each year. Steps include replacing incandescent light bulbs with CFL bulbs, insulating water heaters, using low-flow showerheads, washing clothes in cold water, and planting a native tree.

Global Warming I$ Good for Business

In spring of 2008, Miami was nominated as a finalist for the National City Livability Award, which is awarded by the United States Conference of Mayors. "Given predictions that a large majority of the world's future population will live in urban settings, with the United States currently having nearly 90 percent of its population in cities," said Mayor Diaz, "the single most critical action we can take to help save our planet is to embrace smart growth, to design cities that make sense."

28

Where We Go from Here

I knew I had hit on the heart of this book when I was interviewing a young woman from Massachusetts. She was talking about her company's product in comparison to "old-fashioned solar panels" when it hit me: *Old-fashioned? Solar panels?* Old-fashioned is not a word I would ever have used to describe solar panels. So what changed? As it turns out, everything.

I originally began writing this book to highlight some of the cleantech innovations Americans have made in response to climate change. In spite of its title, I came to a rather stunning realization: This book is not about global warming. It is not even about the new technologies that are being developed. This book is about change.

We are currently undergoing a period of revolutionary change in the world, not just environmentally, but technologically and economically as well. Choosing not to embrace this change is choosing to get left behind while the rest of the world moves on. The real issue, then, is not if we *should* change but how we *will* change. It is our choice, and it is important to understand that *not* making a choice *is* a choice—in this case to stick with the status quo, which may not be our best option.

Whether or not we agree that global warming is a real phenomenon—or to what degree human beings are a contributing factor—virtually everyone agrees that we are not operating our businesses or living our lives in either an environmentally or an economically sustainable manner. We are throwing off waste that is contaminating our air, water, and land resources. At the same time, our reliance on foreign oil is devastating. Our dollar is weak and the strength of our economy is uncertain. We know we can do better. The question is: How?

As individuals, we can start by taking responsibility for our own choices. We can stop looking to others to make change happen for us and welcome the idea that this is a real-life experiment, and we all get to take a part in it. If we choose not to buy a product, eventually companies will stop producing it. By the same token, if we demand a product, there is certain to be an entrepreneur out there somewhere who will realize the profit potential in our demand and start producing what we need.

As businesses, we can make sure we are actually solving a problem, not just chasing a popular notion or a government incentive. Regardless of how "green" we are (or say we are), the businesses today that will most likely be in business tomorrow are those that actually provide high-quality, cost-effective solutions to some of society's most pressing problems.

As a government, we can make sure the "leadership" role we assume is one that is truly sustainable, both economically and environmentally. Fear-based, reactionary policies do more harm than good in the long run; they rob us of our liberties without achieving the results we wanted in the first place. We must make sure the legislation we enact protects the free-market ecosystem as well as the natural environment. Profit is not a four-letter word, and without capital, there are no new ventures.

It is important to reduce waste and increase efficiencies. It is also important to implement interim technologies and to make trade-offs between existing technologies. In the end, however, we are going to have to research, develop, and deploy new technologies and new ways of doing things in order to reach a point where our productivity and our prosperity are no longer tied to foreign fuel consumption or greenhouse gas emissions. Governments cannot mandate innovation; however, they can help lay the foundation from which individuals and businesses can create their own innovations.

Where We Go from Here

Climate change has the potential to force all of us to improve our economic and environmental well-being. However, it is up to each one of us to be an active part of that shift in order to be successful. We know, intuitively, that if we want to do more than just survive the current change in climate, if we want to actually *thrive* in an environment that is fraught with dangers (and opportunities), then we are going to have to do something else. We are going to have to grow and develop and evolve. We are going to do what we do best—come up with *real* solutions to *real* problems that no one else in the world has ever dreamed of. Global warming is not just good for business; it is perhaps one of the greatest catalysts for change that humankind has ever faced. Where we go from here is entirely up to us.

Author's Note

Writing about cutting-edge technologies is one of the most exciting things in the world; but, in the months it has taken to complete this book, the facts and figures that I have tried so painstakingly to collect have been in a constant state of flux, making it difficult if not impossible to stay current on any given technology, let alone the many technologies that are evolving.

Some of the most intriguing projects I have seen have either changed into something else or ceased to exist completely. At the same time, new technologies and new methods for measuring the effectiveness of those technologies are being developed on a daily basis.

I have tried to be as thorough and as accurate as possible, while at the same time rendering a vast amount of complex and rapidly shifting data in an extremely simplified and generalized format. I apologize in advance for any errors or omissions on my part. Hopefully, at least, these stories will entertain and pique people's interests enough that they will go on to learn more and—maybe even more importantly—do more with what they have learned.

Glossary

Active Fuel Management™ (AFM)

General Motors' trademarked name for a variable displacement internal combustion engine technology that reduces fuel consumption by deactivating cylinders, thus enabling a V8 engine to shift from 8 cylinders to 4 cylinders and then back to 8 cylinders, depending on load conditions.

http://activefuelmanagement.com

Alternative Energy

Energy derived from nontraditional or nonfossil fuel sources. Often used interchangeably with renewable energy; however, alternative energy sources may include nuclear power, "clean" coal, compressed natural gas (CNG), liquefied natural gas (LNG), and other technically nonrenewable resources.

www.epa.gov/climatechange/glossary.html

www.mms.gov/offshore/AlternativeEnergy/Definitions.htm

Aquaculture

A term that refers to "the breeding, rearing, and harvesting of plants and animals in all types of water environments, including ponds, rivers, lakes, and the ocean."

http://aquaculture.noaa.gov/welcome.html

Biodiesel

An alternative fuel, produced from renewable resources, such as soybean or algae oil, that contains no petroleum but can be blended with petroleum diesel to produce a biodiesel blend (i.e. B20 fuel).

http://www.biodiesel.org/resources/biodiesel_basics

Bioplastic

Plastic that is made from biorenewable materials such as corn starch.

http://www.global-greenhouse-warming.com/glossary-climate -change.html

Building-Integrated Photovoltaics (BIPV)

The integration of solar photovoltaic (PV) technology into the structural design of a building, replacing traditional construction materials such as roof shingles or window overhangs.

http://www.solar-works.com/solarfacts/glossary

Carbon Footprint

Each of us has an impact on the environment. A carbon footprint is a measure of the impact each of us has in terms of the greenhouse gases we produce in our daily activities.

http://www.carbonfootprint.com

Corporate Average Fuel Economy (CAFE)

CAFE is "the weighted average fuel economy, expressed in miles per gallon (mpg), of a manufacturer's fleet of passenger cars or light trucks with a gross vehicle weight rating (GVWR) of 8,500 lbs. or less, manufactured for sale in the United States, for any given model year." CAFE standards are scheduled to increase to an average of 35 mpg by 2020.

http://www.nhtsa.dot.gov/CARS/rules/CAFE/overview.htm

Carbon Dioxide Capture and Storage (CCS)

Carbon capture and sequestration involves the separation and capture of carbon dioxide from stationary CO_2 sources such as power plants. Carbon may be injected into geologic formations. Plants and microorganisms may also be used in terrestrial applications to store CO_2.

http://www.netl.doe.gov/technologies/carbon_seq/index.html

Clean Technology

Clean technology is often used interchangeably with **green technology**. Cleantech is driven by market economics (versus regulations) and covers several industry sectors, including: energy generation, storage and efficiency, transportation, building and manufacturing, recycling and waste management, and agriculture.

http://cleantech.com

Climate Change

A change in the long-term average weather patterns of a given region, including changes in temperature, cloud cover, precipitation, wind, humidity, and atmospheric pressure.

http://www.epa.gov/climatechange/glossary.html#C

Concentrating Solar Power (CSP)

Solar power plants that collect energy from the sun utilizing various mirror configurations that convert the high-temperature heat into electricity. There are three types of CSP: trough systems, power tower systems, and solar dish/engine systems.

http://www.mtpc.org/cleanenergy/energy/glossarytechfuels.htm

Digester Gas

Digester gas, or anaerobic digester gas (**ADG**), is a biogas that is produced during the decomposition of biomass such as wastewater or manure. During this decomposition process, bacteria digest solid waste in the absence of oxygen (anaerobically) and produce digester gas—a combination of methane and other gases—as a by-product. This biogas can be used as an alternative to natural gas.

http://www.stirlingenergy.com/news-media/glossary.asp

http://www1.eere.energy.gov/femp/newsevents/fempfocus_article.cfm/news_id=8961

Distributed Energy

Distributed energy technologies generate small-scale power to homes, businesses, and other users. These technologies are located close to where electricity is used and can provide either an alternative to or an enhancement of grid-base electric power.

http://www.energy.ca.gov/distgen/index.html

Dual Mode Transportation

A concept wherein vehicles can be transported along automated, high-speed guideways. When they reach their destination, the vehicles can leave the automated guideway and drive on conventional roads.

http://faculty.washington.edu/jbs/itrans/dualmode.htm

Eco-Vehicle Assessment System (Eco-VAS)

Developed by Toyota to allow a systematic assessment of the burden a vehicle will have on the environment as the result of its production, use, and disposal. For every car, six critical dimensions are assessed: fuel efficiency, exhaust emissions, external vehicle noise, reduction of environmental impact, improved recyclability, and reduction of use of substances of concern, such as toxic metals.

http://www.toyota.com/html/hybridsynergyview/2006/summer/ecovas.html

Enhanced Geothermal Systems (EGS)

Enhanced or engineered geothermal systems consist of reservoirs that are "created to produce energy from geothermal resources that are otherwise not economical due to lack of water and/or permeability."

http://www1.eere.energy.gov/geothermal/egs_animation.html

Ethanol 85 (E85)

A motor-fuel blend of 85 percent ethanol and 15 percent gasoline.

http://www.e85fuel.com

Electric Vehicles (EV)

Electric vehicles are an alternative to conventional fossil fuel burning internal combustion engines. EVs use electricity as their fuel source, which allows them to run quietly and without tailpipe emissions.

http://www.eaaev.org

Feed-in Tariff

A feed-in tariff, or feed law, is "a mandated, long-term premium price for renewable energy paid by the local electric utility to energy producers."

http://www.newrules.org

Fission

Very basically, when atoms are split or fissioned into smaller atoms, energy is released. Nuclear fission is used to produce power in over 100 nuclear power plants across the United States.

http://www.inyoyucca.org/glossary.htm

Fuel Cell Vehicle (FCV)

Fuel cell vehicles may be another alternative to fossil fuel burning internal combustion engines. Toyota's Fuel Cell Hybrid Vehicle (**FCHV**) is powered by a fuel cell stack, which gets its energy from high-pressure hydrogen.

http://www.toyota.com/about/our_commitment/environment/vehicles/fuel_cells.html

http://fchv.its.ucdavis.edu

Fusion

Very basically, when atoms are combined or fused together, energy is released. The sun is an example of a nuclear fusion reactor.

http://www.eia.doe.gov/kids/glossary

http://fusedweb.llnl.gov

Geothermal Heat Pump (GHP)

Geothermal heat pumps, or ground source heat pumps (**GSHP**), use the constant temperature of the earth to provide heating, air conditioning, and hot water to a home or business.

http://www.eere.energy.gov/consumer/your_home space_heating _cooling

http://www.energystar.gov

Global Warming

An average increase in the temperature of the Earth's surface atmosphere. Global warming can occur as a result of both natural and anthropogenic (human-caused) activities. Greenhouse gas emissions (i.e. from cars and power plants) are believed to trap heat in the Earth's atmosphere, thus contributing to global warming.

http://www.epa.gov/climatechange/glossary.html#G

Green Collar Jobs

Although there are many definitions, green collar jobs are essentially jobs that involve eco-friendly products and services.

http://www.alternative-energy-news.info/white-blue-green-collar

Greenhouse Gas (GHG)

Greenhouse gases trap the sun's heat in the atmosphere in what is known as the **greenhouse effect**. Major greenhouse gases include carbon dioxide (CO_2), methane (CH_4), nitrous oxide (N_2O), and fluorinated gases (HFCs, PFCs and SF_6).

http://www.eia.doe.gov/bookshelf/brochures/greenhouse/Chapter1.htm

Green Hotel

Green hotels are managed in an eco-friendly manner. For example, green hotels can have water- and energy-saving devices as well as recycling and solid waste disposal programs that increase efficiency, reduce waste (and costs), and thus help the environment.

http://www.greenhotels.com

Greenwashing

Greenwashing is "the act of misleading consumers regarding the environmental practices of a company or the environmental benefits of a product or service."

http://www.terrachoice.com

Hybrid Electric Vehicle (HEV)

Hybrid electric vehicles, or hybrids, have both an internal combustion engine and an electric motor that is charged by a battery, which is in turn recharged by the internal combustion engine. Toyota's Prius is one example of a hybrid.

http://www.nrel.gov/vehiclesandfuels/hev/hevs.html

Hypermiling

Hypermiling is a term used when drivers increase gas mileage by altering the way they drive. For example, accelerating at a slow, even pace, driving the speed limit, and avoiding unnecessary braking can improve your car's gas mileage.

http://www.hypermiling.com

Integrated Gasification Combined Cycle (IGCC)

Integrated gasification combined cycle (IGCC) technology is designed to "clean" heavy fuels such as coal, heavy oil, and pet coke and convert them for use in gas turbines.

http://www.gepower.com/prod_serv/products/gas_turbines_cc/en/igcc/index.htm

Independent System Operator (ISO)

An independent system operator is an entity that monitors the reliability of the power system and coordinates the supply of electricity in a given state or region. For example, the California ISO is "a not-for-profit, public-benefit corporation charged with operating the majority of California's high-voltage wholesale power grid."

http://www.caiso.com

ISO New England meets the electricity needs of New England by "ensuring the day-to-day reliable operation of New England's bulk power generation and transmission system, by overseeing and ensuring the fair administration of the region's wholesale electricity markets, and by managing comprehensive, regional planning processes."

http://www.iso-ne.com

Investment Tax Credit (ITC)

The investment tax credit (ITC) reduces federal income taxes based on capital investment in renewable energy projects. The ITC helps offset up-front costs and provides incentives to develop capital-intensive renewable energy technologies.

http://www.wri.org/publication/bottom-line-series-renewable -energy-tax-credits

Kilowatt (kW)

Electrical power is measured in watts (W), kilowatts (kW), megawatts (MW), gigawatts (GW) and so on. 1,000 watts of electricity equals one kilowatt.

http://www.safeelectricity.org/esw_v1_1/glossary/index.html

Kilowatt Hours (kWh)

Electrical energy is measured in terms of power and time. For example, if you use a 200W bulb for five hours, you have used 1,000 watt hours (Wh) or one kilowatt hour (kWh) of electrical energy.

http://www.eia.doe.gov/kids/energyfacts/sources/electricity. html

Life Cycle Analysis (LCA)

Life cycle analysis, or life cycle assessment, involves the thorough evaluation of the environmental and economic impact that a product has from production to disposal. Also known as well-to-wheels (**WTW**) analysis.

http://www.pbs.org/strangedays/glossary

Leadership in Environment and Energy Design (LEED) Green Building Rating System™

A third-party certification program for the design, construction, and operation of green buildings. LEED has become the nationally accepted benchmark for sustainable building.

http://www.usgbc.org

Magnetic Levitation (Maglev)

A rail technology in which trains are levitated above a magnetized track. Without friction, they are able to travel at high speeds.

http://wordnet.princeton.edu

Megawatt (MW)

1 million watts (W) or 1,000 kilowatts (kW) of electricity. *See kilowatt (kW).*

Nanotechnology

A nanometer is one billionth of a meter. A sheet of paper is 100,000 nanometers thick. Nanoscience involves the discovery of properties and behaviors of materials that are measured on a nanoscale. Nanotechnology involves manipulating, controlling, and putting those scientific discoveries to work.

http://www.nano.gov/html/facts/whatIsNano.html

Nongovernment Organization (NGO)

A voluntary (i.e. non-profit) group that caries out environmental or humanitarian missions, encourages political participation at local levels, and makes governments aware of citizens' concerns.

http://www.socgen.com/csr/sustainable_development/glossary.html

Nonprofit Organization (NPO)

In the United States, a nonprofit or not-for-profit organization may be tax-exempt and is not operated for the purpose of making a profit.

http://bhs.econ.census.gov/econhelp/glossary

Net Metering

Net metering enables those with on-site renewable energy sources (i.e. solar) to turn their electric meters backwards when they generate electricity in excess of demand. This allows them to offset their consumption over a given billing period and receive retail prices for the excess electricity they generate.

http://www.eere.energy.gov/greenpower/markets/netmetering.shtml

Ocean Thermal Energy Conversion (OTEC)

OTEC utilizes the natural temperature gradient—the difference in temperature between the sun-heated surface waters and the cold deep sea waters of the ocean—to produce electrical energy.

http://www.nrel.gov/otec

Personal Rapid Transit (PRT)

A rapid transit concept, akin to the automated people mover (APM), for transporting small groups of people in automated vehicles along elevated guideways.

http://www.cprt.org/CPRT/Home.html

Plug-in Hybrid Electric Vehicle (PHEV)

A plug-in hybrid electric vehicle can run on either electricity or gasoline. The batteries in a PHEV can be charged from a standard 120-volt outlet and used to power the vehicle for short trips. The internal combustion engine can operate for longer trips, giving the vehicle a longer range than the standard EV.

http://www.calcars.org/vehicles.html

Production Tax Credit (PTC)

The production tax credit reduces the federal income taxes of qualified owners of renewable energy projects based on the electrical output of their facilities. Each kilowatt-hour (kWh) generated and supplied to the electricity grid reduces the amount of federal income tax owed.

http://www.wri.org/publication/bottom-line-series-renewable
-energy-tax-credits

Photovoltaics (PV)

A photovoltaic, or solar electric, system converts the sun's light directly into electricity.

http://www1.eere.energy.gov/solar/solar_glossary.html

Renewable Energy

Renewable energy comes from sources that are inexhaustible, such as solar, wind, geothermal, and hydroelectric.

http://www.nrel.gov/learning/re_basics.html

Renewable Portfolio Standard

A renewable portfolio standard, also known as a renewable energy standard, is "a state policy that requires electricity providers to obtain a minimum percentage of their power from renewable energy resources by a certain date."

http://www.eere.energy.gov/states/maps/renewable_portfolio
_states.cfm

Smog

Smog is a combination of smoke (i.e. from burning coal) and fog. **Photochemical smog** is a photochemical reaction between sunlight and atmospheric pollutants that often forms a hazardous brown haze.

http://www.wisegeek.com/what-is-smog.htm

Currently, the U.S. Environmental Protection Agency has established National Ambient Air Quality Standards (NAAQS) to measure six air contaminants: carbon monoxide (CO), lead (Pb), nitrogen dioxide (NO_2), ozone (O_3), particulate matter (PM), and sulfur dioxide (SO_2).

http://www.epa.gov/air/criteria.html

Solar Thermal

Solar thermal electric systems (such as CSP) convert solar energy to electricity by heating a transfer fluid which, in turn, powers a turbine that drives a generator to provide electricity. Simple low-temperature solar thermal collectors may be used to heat water for homes.

http://www1.eere.energy.gov/solar/solar_glossary.html

Sustainable Seafood

Sustainable seafood, whether fished or farmed, comes from sources that "can maintain or increase production in the long-term without jeopardizing the affected ecosystems."

http://www.mbayaq.org/cr/SeafoodWatch.asp

Triple Bottom Line

The economic, social, and environmental performance of a company.

http://www.getsustainable.net

Zero Energy Home (ZEH)

A zero energy home, or net-zero energy home (**N-ZEH**), is designed and constructed to produce at least as much energy as it consumes annually.

http://www.sustainabilitydictionary.com/n/netzero_energy _home.php

Zero Emission Vehicles (ZEV)

Zero emission vehicles produce zero emissions or pollution.
http://www.driveclean.ca.gov/glossary.php

1-Sun Flash Test

A metric by which the efficiency of solar cells and modules are measured.

http://www.nrel.gov/pv/measurements/current_vs_voltage.html

Bibliography

Articles & Publications:

Bordon, Mark, Jeff Chu, Charles Fishman, Michael A. Prospero, and Danielle Sacks. "Fifty Ways to Green Your Business (and Boost Your Bottom Line)." *Fast Company*, November 2007, 88–99.

Bourne, Joel K. "Green Dreams," *National Geographic*, October 2007, 41–59.

Department of Energy. "Highlighting High Performance." *4 Times Square New York City*. DOE/GO-102001-1352, November 2001.

Department of Energy. "Hydrogen & Our Energy Future." DOE/EE-0320, 1–43.

Department of Energy—National Renewable Energy Laboratory. "Powered by the Sun: 2007 Solar Decathlon." DOE/GO-102007-2459, September 2007.

Elgin, Ben. "Little Green Lies," *Businessweek*, October 29, 2007, 45–52.

Energy Efficiency and Renewable Energy. *Energy Saver$—Tips on Saving Energy & Money at Home*, September 2007, 1-33. www.eere.energy.gov/consumers/tips

Energy Information Administration. "Annual Energy Review 2006."DOE/EIA-0384 (2006), June 2007, 1–441.

Energy Information Administration. *International Energy Outlook 2006*, DOE/EIA-0484 (2007), May 2007, 1–10.

Home Depot. "Solar Power Systems Installed for You," Home Services brochure, June 27, 2007.

Madison County Farm Bureau. "Illinois Trivia," *Madison County Farm Bureau News,* March 2008, 4.

Morris, Betsy. "What Makes Pepsi Great?" *Fortune*, Volume 157, Number 4, March 3, 2008, 54–66.

Pacala, S., and R. Socolow. "Stabilization Wedges: Solving the Climate Problem for the Next 50 Years with Current Technologies." *Science*, Vol. 305, No. 5685, August 13, 2004, 968–972.

Southwest Windpower, Inc. "Introduction—Skystream 3.7," company information sheet, 1–3.

Southwest Windpower, Inc. "Residential Wind and Your Neighborhood," company information sheet, 1–2.

Southwest Windpower, Inc. "Skystream Early Adopters Achieve Serious Savings," company information sheet.

Svoboda, Elizabeth. "Fueling the Future," *Fast Company*, February 2008, 45–47.

Te Brake, William H. "Air Pollution and Fuel Crises in Pre-industrial London, 1250–1650." *Technology and Culture,* Vol. 16, No. 3, July 1975, 337–358.

Tollefson, Jeff. "Not Your Father's Biofuels," *Nature Publishing Group*, 2008, 880–883.

University of California, Davis Center for Entrepreneurship. "Green Technology Entrepreneurship Academy," brochure, 2007.

Interviews & Correspondence:

Adams, Alex—project manager for Miami 21, interview with author, May 27, 2008.

Amundson, Bruce—spokesperson for Weyerhaeuser, phone interview with author, February 11, 2008.

Anderson, Susan—director of sustainable development for the city of Portland, phone interview with author, December 28, 2007.

Auburn, Jill—national program leader for sustainable agriculture within the United States Department of Agriculture Cooperative State Research, Education and Extension Service), phone interview with author, November 19, 2007.

Baird, Ron—executive director of NELHA, phone interview with author, January 29, 2008.

Baxter, Alan Dr.—General Atomics GT-MHR developer, e-mail to author, February 29, 2008.

Berry, David—spokesperson for Flagship Ventures, e-mail with author, May 29, 2008.

Brekken, Ted Dr.—codirector of Oregon State University's Wallace Energy Systems and Renewables Facility, phone interview with author, November 20, 2007.

Campbell, Mark—chairman of Watersaver Technologies, phone interview with author, February 19, 2008.

Carr, David—assistant director of the West Texas A & M Alternative Energy Institute, phone interview with author, November 21, 2007.

Clark, Ed—communication director for Austin Energy, phone interview with author, November 17, 2007.

Frederick, James Dr.—director of the Georgia Institute of Technology's Institute of Paper Science and Technology, phone interview with author, October 31, 2007.

Fouquet, Doug—General Atomics spokesperson, phone interview with author, February 22, 2008.

Garas, Dahlia (University of California, Davis Institute of Transportation Studies Plug-In Hybrid Center), phone interview with author, November 6, 2007.

Gardner, Amy—AIA, LEED and associate professor at the University of Maryland School of Architecture, Planning and Preservation, phone interview with author, November 15, 2007.

Gurol, Sam—director of Maglev systems at General Atomics, e-mails to author, February 20-21, 2008.

Hanis, Monique (Solar Energy Industries Association), e-mail to author, January 14, 2007.

Bibliography

Henderson, Kirston—president of MegaRail, phone interview with author, February 15, 2008.

Jacoby, Bill—president of AeroCity LLC, phone interview with author, April 21, 2008.

Kaleikini, Mike—Puna Geothermal Venture plant manager, phone interview with author, February 8, 2008.

Kearney, Ken (Water Security Corp.), phone interview with author, February 22, 2008.

Keilbach, Robin—owner Keilbach Farms, phone interview with author, January 17, 2007.

Kelly, John—executive director of the Geothermal Heat Pump Consortium, phone interview and e-mail with author, April 16, 2007.

Keseric, Linda (Prospect New Town Colorado), phone interview with author, December 14, 2007.

Klinski, Jo (LEVX), phone interview with author, February 15, 2008.

Kwong, Bill—spokesperson for Toyota, phone interview with author, February 25, 2008.

Lee, Rick—General Atomics DIII-D Tokamak Operations Fusion Education manager, e-mails to author, March 17, 2008.

Lewandowski, Rick—president and CEO of Prism Solar Technologies, Inc., phone interview with author, April 21, 2008.

Lewis, Malcolm Dr.—president of Constructive Technologies Group, phone interview with author, May 20, 2008.

McCann-Gates, Mary—spokesperson for Clipper Wind, phone interview with author, February 26, 2008.

Michaels, Anthony Dr.—director of the University of Southern California Wrigley Institute for Environmental Studies, on-site interview with author, October 11, 2007.

Oster, Daryl (ET3), phone interview with author, February 14, 2008.

Pal, Gregory—senior director of corporate development for LS9, email with author, May 27, 2008.

Perdon, Albert, executive director of the Orangeline Development Authority, phone interview with author, February 26, 2008.

Perkins, Chris—president of Unimodal Systems, CalIT2 presentation, November 13, 2007, and phone interview with author, November 19, 2007.

Perry, Oliver—president of the Eastern Electric Vehicle Club, phone interview with author, October 30, 2007.

Poss, Jim—founder and president of Big Belly Solar, phone interview with author, February 18, 2008.

Reinert, Bill—national manager for Toyota Advanced Technology, CalIT2 presentation, November 13, 2007, and phone interview with author, November 25, 2008.

Rettig, Tracey—spokesperson for United Solar Ovonic, e-mail to author, March 25–26, 2008.

Bibliography

Reynolds, Tom—president of Watersaver Tech, phone interview with author, February 19, 2008.

Ricker, Charles—senior vice president of marketing and business development for BrightSource Energy, phone interview with author, February 22, 2008.

Rodgers, Mark—communications director for Cape Wind, phone interview with author, January 30, 2008.

Ruby, Brian—spokesperson for Metabolix, phone interview with author, February 15, 2008.

Saltman, David (Open Energy Corporation), phone interview with author, February 12, 2008.

Samuelsen, Scott Dr.—director of the National Fuel Cell Research Center at University of California, Irvine, on-site interview with author, December 12, 2007.

Sanders, Mark—chairman of WaterSaver Technologies, phone interview with author, February 19, 2008.

Sauer, Chris—president and CEO of Ocean Renewable Power Company, phone interview with author, February 22, 2008.

Shalev, Eylon Dr.—research scientist for the Seismology Group at Duke University's Division of Earth and Ocean Sciences, phone interview with author, October 18, 2007.

Sharp, Gordon, CEO of Aircuity Inc., phone interview with author, May 20, 2008.

Slevin, Geoff—vice president of sales and marketing for BP Solar in North America, phone interview with author, November 26, 2007.

Starsinic, Nicole—assistant director of the University of California, Davis Center for Entrepreneurship, phone interview with author, November 14, 2007.

Taylor, Dub—director of SECO, phone interview with author, January 2, 2008.

Taylor, William "Trey" H.—president and head of market development for Verdant Power, phone interview with author, January 2, 2008.

Turner, Stephen—managing director of CTG's northeastern office, phone interview with author, May 20, 2008.

Vargas, Maria—director of strategic partnerships within the EPA's Climate Protection Partnership Division. She is also the brand manager for Energy Star, phone interview with author, November 20, 2007.

Walker, Dr. Larry—director of Cornell University's Northeast Sun Grant Institute of Excellence, phone interview with author, October 19, 2007.

Wander, Michelle—director of the University of Illinois Champaign-Urbana Agroecology and Sustainable Agriculture Program, phone interview with author, October 19, 2007.

War, Jan. C.—operations manager at NELHA, e-mails to author, April 21, 2008.

Wemet, Tracy—spokesperson for Konarka, phone interview with author, February 18, 2008.

Wilkie, Jennifer spokesperson for Terracycle Inc., phone interview with author, February 22, 2008.

Zimmerman, Jackie—general manager for Troutlodge, phone interview with author, February 26, 2008.

Movies:

Gore, Al. *An Inconvenient Truth*, 2006.

Online Sources:

Aircuity Inc. "The Aircuity Center for Green Building Technology," brochure, 2007. http://aircuity.com/Marketing/documents/ACGBT%20 Brochure% 205-10-08.pdf

Aircuity Inc. "Corporate Overview," 2007. http://aircuity.com/Marketing/index.asp?p=Corporate_Overview

Alter, Lloyd. "Terracycle and Sponsored Waste," *Treehugger*, February 12, 2008. http://terrcycle.net/yb_press.htm

American Automobile Association. "Daily Fuel Gauge Report," June 1, 2008. http://www.fuelgaugereport.com

American Electric Power. "IGCC plant in West Virginia could be online in 2012 if plan filed by AEP's Appalachian Power gains approvals," news release, June 18, 2007. http://www.aep.com/newsroom/newsreleases/default.asp

American Wind Energy Association. "Resources—Wind Energy Policy Issues," 1996–2007. http://www.awea.org/faq/wwt_policy .html

American Wind Energy Association. "Wind Energy Costs," FAQS, 2007 http://www.awea.org/faq/wwt_costs.html

American Wind Energy Association. "Wind Energy and the Environment," FAQS, 2007. http://www.awea.org/faq/wwt _environment.html

American Wind Energy Association. "Wind Energy Potential," FAQS, 2007. http://www.awea.org/faq/wwt_potential.html

American Wind Energy Association. "Wind Energy 101." http://www.ifnotwind.org/we101/wind-energy-basics.shtml

American Wind Energy Association. *Wind Energy Teacher's Guide*, 2003, 1-15. http://www.awea.org/pubs/documents/ TeachersGuide.pdf

Amtrak. "Travel Green with Amtrak," 2008. http://www.amtrak. com/servlet/ContentServer?pagename=Amtrak/am2Copy/ Title_Image_Copy_Page&cid=1178294035703

Association of American Railroads. *Overview of US Freight Rail- roads*, January 2008, 1-8. http://www.aar.org/PubCommon/ Documents/AboutTheIndustry/Overview.pdf

Austin Energy. *Green Building Program Newsletter*, July 2003. http://www.austinenergy.com/Energy%20Efficiency/Programs/ Green%20Building/Members/newsletter/pdfs/July2003.pdf

Automobile Association of America (AAA). "Daily Fuel Gauge Report." http://www.fuelgaugereport.com

Bibliography

Ballard, Ernesta. "Business Day: UN Climate Change Conference 2007," speech, December 10, 2007. http://www.weyerhaeuser .com/Company/Media/Speeches

Barrie, John. "Spray on Solar Uses Buckyballs," *EcoGeek*, July 20, 2007. http://www.ecogeek.org/content/view/814/83

Biello, David. "Clean Coal Power Plant Canceled—Hydrogen Economy, Too," *Scientific American*, February 6, 2008. http://www.sciam.com/article.cfm?id=clean-coal-power -plant-canceled-hydrogen-economy-too

BigBellySolar. "Facts and figures about BigBelly." http:// bigbellysolar.com/for-media

BigBellySolar. "Green Building." http://bigbellysolar.com/ about-us/environmental-and-green-technology/green -building.html

BIOstock Blog. "Celunol produces Ethanol from Wood using Bacteria," January 20, 2007. http://biostock.blogspot.com/2007/01/ celunol-produces-ethanol-from-wood.html

Bloomberg, Michael R. "Testimony Before the House Select Committee on Energy Independence and Global Warming," November 2, 2007. http://www.usmayors.org/climateprotection/ documents/testimony/20071102-bloomberg.pdf

Bodman, Sam. "Ask the White House," February 23, 2007. http://www.whitehouse.gov/infocus/energy

BP Solar USA. "About BP Solar," BP plc, 1996–2008. http://www.bp.com

BP Solar USA. "BP Solar groundbreaking ceremony to celebrate expansion project," news release, July 16, 2007. http://www .bp.com

BP Solar USA. "How to buy solar," BP plc, 1996–2008. http://www.bp.com

BrightSource Energy. "Ten FAQS About Solar Thermal Power." http://www.brightsourceenergy.com/faq.htm

Broer, Wendel. "Harvesting energy from algae," *Shell World*, February 15, 2008. http://www.shell.com/swonline

Browne, Lord John. "Energy and the Environment, 10 Years On," BP Group speech at Stanford University, California, April 26, 2007. http://www.bp.com/genericarticle .do?categoryId=98&content Id=7032698

Burritt, Chris. "Dow Chemical's Price Increase Paves Way for Hershey (Update 1)," *Bloomberg News*. May 29, 2008. http:// www.bloomberg.com/apps/news?pid=newsarchive&sid=aAmR .NPK90fU

Bush, George W. "Letter from the President to Senators Hagel, Helms, Craig, and Roberts," The White House, March 13, 2001. http://www.whitehouse.gov/news/ Releases/2001/03/20010314.html

Bush, George W. "Protecting our Nation's Environment," The White House, September 28, 2007. http://www .whitehouse.gov/infocus/environment

Bush, George W. "State of the Union Address," The White House. January 31, 3006. http://whitehouse.gov/news/ releases/2006/01/ 20060131-10.html

Bibliography

Byrd-Hagel Resolution, 105th Congress, 1st Session, S.Res.98, July 25, 1997. http://www.nationalcenter.org/KyotoSenate.html

California Energy Commission. "Anaerobic Digestion," April 27, 2005.http://www.energy.ca.gov/development/biomass/anaerobic.html

Calpine Corporation. "About Geothermal Energy." http://www.geysers.com/geothermal.htm

Carbon Tax Center. "FAQS," February 26, 2008. http://www.carbontax.org/faq/#phase-in

Carbon Tax Center, "Introduction," May 23, 2008. http://www.carbontax.org/introduction/#no-tax-increase

Carbon Tax Center. "Why revenue-neutral carbon taxes are essential, what's happening now, and how you can help." http://www.carbontax.org

Carter, Jimmy. "State of the Union Address," The White House, April 18, 1977. http://www.pbs.org/wgbh/amex/carter/filmore/ps_energy.html

Carey, John and Adrienne Carter, "Food vs. Fuel," *BusinessWeek*, February 5, 2007. http://www.businessweek.com/magazine/content/07_06/b4020093.htm

Carlson, Darren K. "Public Warm to Nuclear Power, Cool to Nearby Plants," *gallup.com*, May 3, 2005. http://www.gallup.com/poll/16111/Public-Warm-Nuclear-Power-Cool-Nearby-Plants.aspx

Casey, Michael. "World's Coal Dependency Hits Environment," *Associated Press—USA Today*, November 4, 2007. http://www .usatoday.com/weather/environment/2007-11-04-china-coal_N .htm

Center for Responsible Nanotechnology. "What is Nanotechnology?" http://www.crnano.org/whatis.htm

Center for Small Business and the Environment, "Did You Know That...." http://www.geocities.com/aboutcsbe

Center for Small Business and the Environment, "Our Mission: The Greening of Small Business." http://www.geocities.com/ aboutcsbe/mission/html

Central Japan Railway Company. "Superconducting Maglev," 2008. http://www.jr-central.co.jp/eng.nsf/english/maglev

Chevron Corporation. "What is Geothermal Energy?" March 2008. http: www.chevron.com/deliveringenergy/geothermal

City of Austin. "Mayor Wynn Advances Federal Climate and Energy Agenda At US Conference of Mayors," news release, June 6, 2007. http://www.austintexas.gov/council/wynn_cpp .htm

City of Austin. "Mayor Wynn Announces Action on Zero Energy Homes," news release, October 18, 2007. http: ww.ci.austin.tex .us/news/2007/zech_release.htm

City of Austin. "Wynn Announces Austin Climate Protection Plan," news release, February 7, 2007. http://www.ci.austin .tx.us/council/mw_acpp_release.htm

Bibliography

City of Chicago. "About the Rooftop Garden," *Department of Environment.* http://egov.cityofchicago.org

City of Grand Prairie. "Sun and Wind Power a Street Light in Grand Prairie, Texas," news release, August 22, 2007. http://www.windenergy.com/globalinstalls/global_installs _grandprarie.htm

City of Miami. "Water Conservation." http://www.miamigov.com/ msi/pages/WaterConservation/default.asp

City of Portland. "Green Investment Fund Offers $425,000 for High Performance Buildings," news release, December 1, 2007. http://www.portlandonline.com/ osd/index.cfm?a=176230&c=42298

Clean Energy Alliance. "Our Mission." http://www.cleanenergyalliance.com/mission.php

CleanTech Network. "Who We Are," 2008. http:// cleantechnetwork.com/index.cfm?pageSRC=Cleantech Defined

Clipper Windpower. "About Us." http://clipperwind.com/ profile.html

Clipper Windpower. "Clipper-developed Silver Star Project Breaks Ground in Texas," news release, September 12, 2007. http://www.clipperwind.com/pr_9-091207.html

Clipper Windpower. "Clipper Leads on Technology and Size as It Develops the Britannia Offshore Wind Turbine," news release, October 8, 2007. http://www.clipperwind.com/pr_100807.html

Clipper Windpower. "Clipper Recognized by US Department of Energy for Outstanding Contribution Toward Energy Industry Advancements." http://www.clipperwind.com/pr_091007.html

Conner, Chuck. "Remarks to the South Dakota Corn Growers Association," speech, January 5, 2008. http://www.usda.gov

Cornell News. "Ethanol Fuel from Corn Faulted as 'Unsustainable Subsidized Food Burning'" http://www.news.cornell.edu/releases/Aug01/corn-basedethanol.hts.html

Cowen, Richard. "Chapter Eleven: Coal." Essays on Geology, History, and People. UC Davis Geology 115, class notes. http://www.geology.ucdavis.edu/~cowen/~GEL115

Cox, Vivienne. "The Business Case for Low Carbon Power— An International Perspective," speech to the Indian Institute of Energy, India, March 19, 2007. http://www.bp.com/genericarticle.do?categoryId=98&contentId=7031191

Cox, Vivienne. "Why Clean Electricity is Critical in Combating Climate Change—Steps to Accelerate America's Low Carbon Power Economy," speech, September 27, 2006. http://www.bp.com/genericarticle.do?categoryId=98&contentId=7023515

Davies, John. "Strategic Thinking." *GreenBiz.com*, January 2008. http://greenbiz.com/news/columns_third.cfm?NewsID=36478

Department of Energy, "About DOE." http://www.energy.gov/about/index.htm

Department of Energy, "FutureGen—Tomorrow's Pollution-Free Power Plant." http://fossil.energy.gov/programs/powersystems/futuregen

Bibliography

Department of Energy, "Gen IV Nuclear Energy Systems," Office of Nuclear Energy. http://www.ne.doe.gov/genIV/ neGenIV1.html

Department of Energy. "Hydropower." http://www.energy.gov/ energysources/hydropower.htm

Diaz, Manuel A. "Testimony before the House Select Committee on Energy Independence and Global Warming," speech, November 2, 2007. http://www.usmayors.org/climateprotection/ documents/testimony/20071102-diaz.pdf

Dickerson, Paul. "Remarks to the 20th NREL Industry Growth Forum," November 7, 2007. http://www.eere.energy.gov/news/ speedhes/2007-11-07_20th_igf.cfm

Douglass, Elizabeth and Victoria Kim. "PG&E to Get Watts from Waves." *Los Angeles Times*, December 19, 2007. http://www .finavera.com/files/2007-12-19%20LA%20Times%20 PG&E%20to %20get%20watts%20from%20 waves.pdf

Dow Chemical Company. "Dow Responds to Surging Energy Costs," news release, May 28, 2008. http://news.dow.com/ dow_news/corporate/2008/20080528a.htm

Dow Jones Financial Information Services. "Driven by US Enthusiasm, Global VC Investment in Clean Technologies Jumps 43% in 2007 to $3 Billion," news release, February 29, 2008. http://www.fis.dowjones.com/Y2007CleantechPR.pdf

DriveClean. "What is a TLEV, LEV, ULEV, SULEV, PZEV, AT PZEV and ZEV?" FAQ. http://driveclean.ca.gov/en/gv/faq/ index.asp

DSIRE. "Renewable Energy Production Tax Credit (PTC)," 2007. http://www.dsireusa.org/library/includes/incentive2 .cfm?Incentive_Code=US13F&State=Federal%C2%A4tpageid=1

Duke Energy. "How Do Nuclear Plants Work?" http://www .duke-energy.com/about-energy/generating-electricity/ nuclear.asp

Duke Energy. "Questions and Answers." http://www.duke-energy .com/power-plants/new-generation/edwardsport-faq.asp

Edison International. "Edison's Nuclear Generation Program," *SCE Backgrounder*, December 18, 2007. http://www .edison news.com

Edison International. "The Future of Nuclear Generation," *SCE Backgrounder*, December 10, 2007. http://www .edisonnews.com

Edison International. "San Onofre's Future," *SCE Backgrounder*, December 8, 2007. http://www.edisonnews.com

Edison International. "Southern California Edison." http://www.edison.com/ourcompany/sce.asp

Edison International. "Southern California Edison Corporate Profile," *SCE Backgrounder*, January 25, 2008. http://www .edisonnews.com

Ehrlich, David. "Clipper Windpower to Build World's Largest Turbine," CleanTech Network, October 8, 2007. http://media .cleantech.com/1883/clipper-windpower-to-build-worlds -largest-turbine

Bibliography

Electric Auto Association. "EV Timeline," 2007. http://www.eaaev
.org/history/index.html

Energy Efficiency and Renewable Energy. "Biofuels Initiative,"
Biomass Program, January 22, 2007. http://www1.eere.energy
.gov/biomass/biofuels_initiative.html

Energy Efficiency and Renewable Energy. "About the Program,"
Solar Energy Technologies Program, October 24, 2006.
http://www1.eere.energy.gov/solar/about.html

Energy Efficiency and Renewable Energy. "GE Unveils Hybrid
Locomotive for Freight Trains," *EERE Network News*, May
30, 2007. http://www.eere.eneergy.gov/news/news_detail.cfm/
news_id=10991

Energy Efficiency and Renewable Energy. "A History of Geothermal Energy in the United States," Geothermal Technologies
Program, November 1, 2006. http://www1.eere.energy.gov/
geothermal/history.html

Energy Efficiency and Renewable Energy. "Moving Toward Zero
Energy Homes," DOE/GO-102003-1828, December 2003. http://
www.eere.energy.gov/buildings/info/documents/pdfs/35317.pdf

Energy Efficiency and Renewable Energy. "Net Metering
Policies," The Green Power Network. http://www.eere.energy
.gov/greenpower/markets/netmetering.shtml

Energy Efficiency and Renewable Energy. "Ocean Thermal
Energy Conversion." http://www.eere.energy.gov/consumer/
renewable_energy/ocean/index.cfm/mytopic=50010

Energy Efficiency and Renewable Energy. "How Hydropower Works," Wind & Hydropower Technologies Program. http://www1.eere.energy.gov/windandhydro

Energy Efficiency and Renewable Energy. "Public Power Wind Pioneer Awards," Wind and Hydropower Technologies Program.http://www.eere.energy.gov/windandhydro/windpoweringamerica/wpa_awards.asp

Energy Efficiency and Renewable Energy. "States with Renewable Portfolio Standards," February 8, 2008. http://www.eere.energy.gov/states/maps/renewable_portfolio_states.cfm

Energy Information Administration. "Biomass Timeline." http://www.eia.doe.gov/kids/history/timelines/biomass.html

Energy Information Administration. "Ethanol Timeline." http://www.eia.doe.gov/kids/history/timelines/ethanol.html

Energy Information Administration. "Geothermal Energy—Energy from the Earth's Core." http://www.eia.doe.gov/kids/energyfacts/sources/renewable/geothermal.html

Energy Information Administration. "US Nuclear Plants: San Onofre."http://www.eia.doe.gov/cneaf/nuclear/page/at_a_glance/reactors/sanonofre.html

Energy Information Administration. "US Nuclear Reactors." http://www.eia.doe.gov/cneaf/nuclear/page/nuc_reators/reactsum.html

Energy Information Administration. "Wind Energy—Energy from Moving Air." http://www.eia.doe.gov/kids/energyfacts/sources/renewable/wind.html

Bibliography

Energy Information Administration—National Energy Information Center. "What are the Sources of Greenhouse Gases?" *Greenhouse Gases, Climate Change, and Energy*, April 2, 2004. http://www.eia.doe.gov/oiaf/1605/ggccebro/chapter1.html

EnergyStar. "EnergyStar Qualified New Homes." http://energystar.gov/index.cfm?c=new_homes.hm_index

EnergyStar. "Find Expert Help." http://www.energystar.gov/index.cfm?c=expert_help.find_exp_help

EnergyStar. "Guidelines for Energy Management." http://www.energystar.gov/index.cfm?c=guidelines.guidelines_index

EnergyStar. "Home Energy Audits." http://energystar.gov/index.cfm?c=home_improvement.hm_improvement_audits

EnergyStar. "Portfolio Manager." https://www.energystar.gov/istar/pmpam

EnergyStar. "Seal Not, Save Not." http://www.energystar.gov/ia/home_improvement/home_sealing_article_sample.pdf

EnergyStar. "Special Salute: The First 100 Energy Star Buildings," *Energy & Environmental Management*, 19–21. http://www.energystar.gov/ia/business/Four_Times_Square.pdf

EnergyStar. "Where Does My Money Go?" http://www.energystar.gov/index.cfm?c=products.pr_pie

Environmental Defense Fund. "Making Smart Choices When Eating Seafood," *Seafood Selector*, 2008. http://www.edf.org/page.cfm?tagID=1521

Environmental Protection Agency. "Climate Change: What Is It?"
http://epa.gov/climatechange/kids

Environmental Protection Agency. "Climate Leaders."
http://www.epa.gov/climateleaders

Environmental Protection Agency, "Methane," October 19, 2006.
http://epa.gov/methane/scientific.html

Environmental Protection Agency. "National Top 25: As of January 8, 2008." http://epa.gov/greenpower/toplists/top25.htm

Environmental Protection Agency. "National Ambient Air
Quality Standards (NAAQS)," Technology Transfer Network,
September 26, 2007. http://www.epa.gov/ttn/naaqs

Environmental Protection Agency. "Nitrogen Oxides." http://www
.epa.gov/OMS/invntory/overview/pollutants/nox.htm

Environmental Protection Agency. "Personal Emissions
Calculator." http://www.epa.gov/climatechange/emissions/
ind_calculator.html

Environmental Protection Agency. "Draft Inventory of US Greenhouse Gas Emissions and Sinks: 1990–2006," February 2008,
1-40.http://epa.gov/climatechange/emissions/downloads/08
_Agricul ture.pdf

Extengine. "Diesel Emission Control." http://www.extengine.com/
diesel.php?osCsid=3730e91c18e4ebced5d84bcbcc6190b2

Fairly, Peter. "Does Fusion Have a Future?" *IEEE Spectrum* magazine, February 14, 2008. http://spectrum.ieee.org/feb08/5980

Bibliography

Farrar, Lara. "How to Harvest Solar Power? Beam It Down from Space!" 2008. http://www.cnn.com/2008/TECH/science/05/30/space.solar/index.html

Farrell, John. "Minnesota Feed-In Tariff Could Lower Cost, Boost Renewables and Expand Local Ownership," New Rules Project, January 2008. http://www.newrules.org/de/feedin.html

Fleming, Kent. "Moi Way," Slow Food International, February 10, 2006. http://www.slowfood.com/sloweb/eng/dettaglio.lasso?cod=5232 CF8A0c41e2767Avjg2821112

FutureGen Alliance. "FutureGen Alliance Board Unanimously Agrees that FutureGen at Mattoon Remains in the Public Interest," February 7, 2008. http://www.futuregenalliance.org/news/releases/pr_02-07-08.stm

FutureGen Alliance. "FutureGen Technology," December 2007. http://www.futuregenalliance.org/technology.stm

Gaul, Alma. "Into the Wind: Q-C Area Man Makes His Own Electricity," *Quad City Times*, January 25, 2008. http://www.windenergy.com/news/news_QuadCityTimes_ 1-25-08.html

General Electric Energy. "IGCC History," 1997–2008. http://www.gepower.com/prod_serv/products/gas_turbines_cc/en/igcc/history.htm

General Atomics. "Agreement Signed for ITER International Fusion Energy Project," news release, November 22, 2006. http://www.ga.com/news.php

General Atomics. "Entergy's Keuter Elected Chairman of Advanced Reactor Advisory Board," news release, November 7, 2001. http://www.ga.com/news.php

General Atomics. "What Is Fusion?" http://fusion.gat.com/global/
Basics

General Electric. "Hybrid Locomotive: the future of rail is just around the corner," Ecomagination fact sheet, 2005, 1–2. http://ge.ecomagination.com/site/downloads/hybr/Hybrid _onepager_en. pdf

General Electric. "GE Unit Partners with SunPower on California Solar Projects," news release, January 8, 2007. http://www.genewscenter.com/Content/Detail.asp?ReleaseID =2920&NewsAreaID=2&MenuSearch CategoryID=5

General Motors. "2005/06 Corporate Responsibility Report." http://www.gm.com/corporate/responsibility/reports/06/400 _products/1_ten/410.html

General Motors. "Chevrolet's Project Driveway Fuel Cell Drivers Get Convenient Fuel at New York's First Hydrogen Station," news release, November 13, 2007. http://media.gm.com/news/ press/pr_recent/index.html

General Motors. "Chevrolet Tahoe with Two-Mode Full Hybrid Exploits Fuel Savings, SUV Driving Pleasure and Perfor-mance," news release, January 9, 2006. http://www.gm.com/ experience/technology/news/2006/hybrid_tahoe_010906.jsp

General Motors. "GM Extends Biofuels Leadership with Coskata Partnership," news release, January 13, 2008. http://media .gm.com/news/press/pr_recent/index.html

GeoComfort. "Advantages of Geothermal," 2008. http://geocomfort .com/?page=Geothermal_Technology/dvantages_of_Geothermal

Bibliography

Geo Source One. "Four Times The Efficiency." http://www
.geosource one.com/wisg.php

Geothermal Energy Association. "All About Geothermal Energy—
Current Use." http://www.geo-energy.org/aboutGE/currentUse
.asp

Geothermal Energy Association. "What Are the Different
Ways in Which Geothermal Energy Can Be Used?" 2008.
http://www.renewableenergyaccess.com/rea/partner?cid=4102

Geothermal Heat Pump Consortium. Press Kits. Contractors.
2003. http://geoexchange.us/press/contractors.htm

Giller, Chip. "The Way I See It #289," Starbucks. http://www
.starbucks.com/retail/thewayiseeit_default.asp?

Global Nuclear Energy Partnership. "DOE Statement on Canada
Joining the Global Nuclear Energy Partnership," news release,
November 30, 2007. http://www.gnep.energy.gov/gnepPRs/
gnepPR113007.html

Global Nuclear Energy Partnership. "Greater Energy Security
in a Cleaner Safer World." http://www.gnep.energy.gov/
gnepProgram.html

Global Warming. Dictionary.com. *The American Heritage® New
Dictionary of Cultural Literacy, Third Edition.* Houghton
Mifflin Company, 2005. http://dictionary.reference.com/browse/
global warming

Grand Coulee Dam. "Hydroelectric Power Generation."
http://users.owt.comchubbard/gcdam/html/hydro.html

GreenFuel Technologies Corporation. "FAQS." http://www
.green fuelonline.com/contact_faq.html

Greer, Diane. "Creating Cellulosic Ethanol: Spinning Straw into
Fuel." *BioCycle* eNews Bulletin, May 2005. http://www.harvest
cleanenergy.org/enews/enews_0505/enews_0505_Cellulosic
_Ethanol.htm

Goldweber, Aaron. "Q&A: Larry Walker Calls for a 'Manhattan
Project' for Energy in Biofuels," *ChronicleOnline*, February 23,
2006. http://www.new.cornell.edu/stories/Feb06/Larry_Walker
_QA.aw.html

Hargadon, Andrew. "The Energy Revolution, in Perspective,"
blog, November 16, 2007. http://andrewhargadon.typepad.com/
my_weblog/2007/11/index.html

Harvey, Fiona. "A Good Time to Be a Green Entrepreneur,"
Financial Times: Business Life—Enterprise, February 22,
2006. http://www.newenergyfinance.com/NEF/HTML/
PressCoverage/2006-02-22_FT_Green.pdf

Hayward, Tony. "Energy Security and America," speech at the
Houston Forum, November 8, 2007. http://www.bp.com/
genericarticle.do?category Id=98&contentId=7038285

Hirsch, E.D. Jr., Joseph F. Kett, and James Trefil. "Investment
Tax Credit." *The New Dictionary of Cultural Literacy, Third
Edition*, Houghton Mifflin Company, 2002. http://www
.bartleby.com/ 59/18/investmentta.html

Heslin Rothenberg Farley & Mesiti P.C. "Clean Energy Patent
Growth Index." http://cepgi.typepad.com

Bibliography

Honda. "Honda Introduces Experimental Home Energy Station IV," November 17, 2007. http://world.honda.com/news/2007/ 4071114Experimental-Home-Energy-Station

HR Biopetroleum. "FAQS." http://www.hrbiopetroluem.com/html/ faq.html

HR Biopetroleum. "The Next Generation of Biofuels." http://www.hrbiopetroleum.com/index.html

IBM. "IBM Pioneers Process to Turn Waste into Solar Energy," news release, October 20, 2007. http://www-03.ibm.com/ press/us/en/pressrelease/22504.wss

Inc.com. "OfficeMax, TerraCycle Offer Green Office Products," *Inc.com*, May 8, 2008. http://www.inc.com/news/ articles/2008/05/08office.htm

Inc. Staff. "The Road Crew, "*Inc. Magazine*, November 2006. http://www.inc.com/magazine/20061101/green50_intro.html

Independent News and Media. "TGV Lives Up to Its Name with 357mph Record," April 4, 2007. http://www.independent.co.uk/ news/world/europe/tgv-lives-up-to-its-name-with-357mph -record-443244.html

Innovative Transportation Technologies. "Dualmode Transportation Concepts," June 1, 2008. http://faculty.washington.edu/jbs/ itrans/dualdbte.htm

Innovative Transportation Technologies. "Personal Rapid Transit (PRT) or Personal Automated Transport Quicklinks," January 10, 2008. http://faculty.washington.edu/jbs/itrans/prtquick.htm

International Ground Source Heat Pump Association. "What Is a Ground Source Heat Pump?" 2006. http://www.igshpa.okstate .edu/geothermal/geothermal.htm

International Herald Tribune. "Oil Surges on Bullish Goldman Sachs Estimate, Bloomberg News. May 16, 2008. http://www .iht.com/articles/2008/05/16/business/16crude.php

Iowa Heat Pump Association. "Geo Exchange Heat Pumps." http://www.iaheatpump.org/iaheatpump/GeoPumps.aspx

Japan Transport Promotion Association. "New Bullet Trains," *Land Vehicles*. http://www.transport-pf.or.jp/english/land/rail/ shinkansen.html

Japanese Lifestyle. "Shinkansen History," February 9, 2008. http://www.japaneselifestyle.com.au/travel/shinkansen _history.htm

Johnson, Stephen L. "Ask EPA," November 1, 2007. http://www.epa.gov/askepa/transcripts/askepa110107.html

Johnson, Stephen. "EnergyStar Change a Light, Change the World But Tour," October 3, 2007 speech. http://yosemite.epa .gov/opa/admpress.nsf

Karsner, Andy. "Ask the White House," March 27, 2007. http://www.whitehouse.gov/infocus/energy

Kenney, Brad. "GM Hybrid-Powered Buses Get Seattle Transit Contract," *IndustryWeek*, May 17, 2007. http://www .industryweek.com/ReadArticle.aspx?ArticleID=14211

Kho, Jennifer. "An iPod for Wind Power?" *Red Herring*, April 24, 2006. http://www.redherring.com/Home/16623

Bibliography

Komanoff, Charles. "Don't Trade Carbon, Tax It." *Grist Magazine, Inc,* February 13, 2007. http://gristmill.grist.org/story/2007/2/13/83257/5462

Komanoff, Charles. "Strange Bedfellows in Climate Politics." *Grist Magazine, Inc,* May 22, 2007. http://gristmill.grist.org/story/ 2007/5/22/7926/58739

Konarka. "Konarka and SKYShades Announce Development Agreement to Integrate Organic Photovoltaic (OPV) Material into Tension Fabric Material," press release, January 30, 2008. http://www.konarka.com/index.php/site/newsdetail

Konarka. "Products." http://konarka.com/products

Kotas, Jerry. *Renewable Energy Credits Another Option in Your Renewable Energy Portfolio,* presentation, July 30–August 1, 2001, 1–10. http://www.eere.energy.gov/greenpower/conference/6gpmc01/jkotas01. pdf

Lawrence Livermore National Laboratory. *Fusion Energy Education.* http://fusedweb.llnl.gov

Lawrence Livermore National Laboratory. https://lasers.llnl.gov

Lingle, Linda. *Hawaii State of the State Address,* January 22, 2008. http://hawaii.gov/gov/leg/2008-session/state-of-the-state/STATE%20OF%20THE%20STATE%20ADDRESS%202008.pdf

LS9. "FAQ." http://www.ls9.com/news/FAQ.html

MacLeod, Mark. "How Does Cap and Trade Work?" *Climate 411,* June 4, 2007. http://environmentaldefenseblogs.org/climate411/2007/06/04

Maryland Department of the Environment. "Did You Know? Fun Facts and Statistics." http://www.mde.state.md.us/Programs/ LandPrograms/Recycling/Education/did_you_know.asp

Massachusetts Institute of Technology. *The Future of Geothermal Energy: Impact of Enhanced Geothermal Systems [EGS] on the United States in the 21st Century*, 2006, 1–372. http://www1 .eere.energy.gov/geothermal/egs_technology.html

McDonough, William, and Michael Braungart. "The NEXT Industrial Revolution," *The Atlantic*, October 1998. http:// www.ratical.org/co-globalize/nextIndusRev.html

Metabolix. "Biotechnology Foundation: Plants," 2006. http://www .metabolix.com/biotechnology%20foundation/plants.html

Metabolix. "Metabolix and ADM to Produce Mirel™, the World's First Biobased and Fully Biodegradable Plastic," news release, April 22, 2007. http://www.metabolix.com/publications/ pressreleases/PR20070423.html

Metabolix. "Our Core Technology." http://www.metabolix.com

Miami21. "Frequently Asked Questions," City of Miami Planning Department, 2005. http://miami21.org/PDFs/ miami21factsenglish.pdf

Miami Dade Transit. "Metromover Information." http://www .co.miami-dade.fl.us/transit/metromover.asp

Monorail Society. "Shanghai, China." http://www.monorails .org/tmspages/MagShang.html

Bibliography

Monterey Bay Aquarium. "Seafood Watch—Seafood Guide," 1999–2008. http://www.montereybayaquarium.org/cr/ SeafoodWatch/web/sfw_faq.aspx

Morelli, Mark. "Business Overview," ECD Annual Meeting of Stockholders, December 11, 2007, 2–18. http://www.ovonic .com/ir_annual_meeting_new.cfm

Moore, Patrick. "Going Nuclear," *Washington Post*, April 16, 2006, B01. http://www.washingtonpost.com/wp-dyn/content/ article/2006/04/14/AR2006041401209.html

National Academies. *Understanding and Responding to Climate Change: Highlights of National Academies Reports*, March 2006, 1–24. http://dels.nas.edu/basc/Climate-HIGH.pdf

National Biodiesel Board. "What is Biodiesel?" http://www .biodiesel.org/pdf_files/fuelfactsheets/kids_sheet.pdf

National Energy Laboratory of Hawaii Authority (NELHA). "NELHA History." http://www.nelha.org/about/history.html

National Energy Laboratory of Hawaii Authority (NELHA). "NELHA Tenants." http://www.nelha.org/tenants/tenants.htm

National Energy Laboratory of Hawaii Authority (NELHA). "Unique Resources." http://www.helha.org/about/resources .html

National Energy Technology Laboratory. "Carbon Sequestration FAQ Information Portal." http://www.netl.doe.gov/technologies/ carbon_seq/FAQs/tech-status.html

National Energy Technology Laboratory. "DOE Announces Restructured FutureGen Approach to Demonstrate Carbon Capture and Storage Technology at Multiple Clean Coal Plants," news release, January 30, 2008. http://www.netl.doe .gov/Publications/press/2008/08003-DOE_Announces_ Restructured_FutureG.html

National Oceanographic and Atmospheric Administration. "China Fisheries," China-US Project, December 2000. http://www.lib .noaa.gov/china

National Renewable Energy Laboratory. "Building-Integrated PV." http://www.nrel.gov/pv/building_integrated_pv.html

National Renewable Energy Laboratory. "Clean Energy Industry Growth Forums." http://www.nrel.gov/technologytransfer/ entre preneurs/ce_growth.html

National Renewable Energy Laboratory. "Clean Energy Investors Directory." http://www.nrel.gov/technologytransfer/ entrepreneurs/directory2.html#3

National Renewable Energy Laboratory. "NREL Overview," July 12, 2007. http://www.nrel.gov/overview

National Renewable Energy Laboratory. "NREL Ranks Leading Utility Green Power Programs," news release, April 3, 2007. http://www.nrel.gov/news/press/2007/506.html

National Renewable Energy Laboratory. "20th NREL Industry Growth Forum." http://www.cleanenergyforum.com

Bibliography

National Renewable Energy Laboratory. The 20th NREL Industry
Growth Forum, forum program, November 6–8, 1–93.
http://www.nrel.gov/technologytransfer/pdfs/growth_forum
_program.pdf

National Renewable Energy Laboratory. "Wakonda Technologies
is the Clean Energy Entrepreneur of the Year," news release,
November 8, 2007. http://www.nrel.gov/news/press/2007/539
.html

National Renewable Energy Laboratory. The Geothermal Technologies Program. "About Geothermal Electricity." http://www
.nrel.gov/geothermal/geoelectricity.html

National Security Space Office. "Space-based Solar Power
as an Opportunity for Strategic Security," October 10,
2007. http://www.nss.org/settlement/ssp/library/final
-sbsp-interim-assessment-release-01.pdf

National Venture Capital Association. "CleanTech Venture
Capital Investments by US Firms Breaks Record in 2007."
http://nvca.org/pdf/CleanTechInterimPR.pdf

New Jersey Institute of Technology. "NJIT Researchers
Develop Inexpensive, Easy Process to Produce Solar Panels,"
news release, July 18, 2007. http://www.njit.edu/news/2007/
2007-265.php

New Resource Bank. "Smart Financing Options for a Smart
Decision." http://www.newresourcebank.com/personal-banking/
loans-and-financing.php#solarhome

Nickels, Greg. "Bright Lights of the Cities: Pathways to a Clean Energy Future," testimony before the House Select Committee on Energy Independence and Global Warming, November 2, 2007. http://www.usmayors.org/climateprotection/documents/testimony/20071102-nickels.pdf

Nuclear Energy Institute. "Life-Cycle Emissions Analysis," June 7, 2007. http://nei.org/keyissues/protectingtheenvironment/lifecycleemissionsanalysis

Nuclear Regulatory Commission. "Fact Sheet on the Three Mile Island Accident," February 20, 2007. http://www.nrc.gov/reading-rm/doc-collections/fact-sheets/3mile-isle.html

Nuclear Regulatory Commission. "Find Operating Nuclear Power Reactors by Location or Name," February 14, 2008. http://www.nrc.gov/info-finder/reactor

Nuclear Regulatory Commission. "Nuclear Reactors," April 20, 2007. http://www.nrc.gov/reading-rm/basic-ref/students/reactors.html

Office of Civilian Radioactive Waste Management. "Yucca Mountain Repository," February 5, 2008. http://www.ocrwm.doe.gov/ym_repository/index.shtml

Office of Management and Budget. "Department of Energy 2008 Budget." http://www.whitehouse.gov/omb/budget/fy2008/energy.html

Open Energy Corporation. "SunCone CSP." http://www.openenergycorp.com/company/product_overview.php

Bibliography

Ormat Technologies, Inc. "Geothermal Development in Utah: The Reality of Today, and the Promise of Tomorrow," 1–22. http://www.utahcleanenergy.org/documents/ORMAT _UtahPolicyPresentation.pdf

Ormat Technologies Inc. "Ormat Desert Peak Aims to be First U.S. Commercial Power Project Using EGS," press release, February 14, 2008. http://www.ormat.com/print.hph?did137

Peterson, Britt. "The Future of Fusion," *Seed* magazine, June 22, 2006. http://seedmagazine.com/news/2006/06/the_future_of _fusion.php

Piquepaille, Roland. "Nanotechnology Used to Study Environment," *Roland Piquepaille's Technology Trends,* January 20, 2005. http://www.primidi.com/2005/01/20.html

Potter, Tom. "State of the City," January 18, 2008 speech. http://www.portlandonline.com/mayor/index.cfm?c=41148&a =181603

Power, Stephen, Rebecca Smith and Jeffrey Ball. "U.S. Drops Coal Project," *The Wall Street Journal*, January 31, 2008. http://online.wsj.com/article/SB12017539758831345.html

Press, Jim, "Toyota Sets New Targets for Improvement in 2006 North American Environmental Report," news release, December 21, 2006. http://pressroom.toyota.com/Releases/ View?id=TYT2006122 176129

Quain, John. "Super Trains: Plans to Fix US Rail Could End Road and Sky Gridlock," *Popular Mechanics*, December 2007. http://www.popularmechanics.com/technology/transportation/ 4232548.html

Rafter, Dan. "Onsite Energy for Affordable Housing," *Distributed Energy: The Journal for Onsite Power Solutions*, September/October 2007. http://www.forester.net/de_0709_onsite.html

RailEurope. "In France, the New SNCF TGV East train Has Broken the World Speed Record," www.raileurope.com/us/rail/tgvest/index.htm

RailPower. "GG Series Yard Switchers," 2004–2007. http://www.railpower.com/products_gg.html

Residential Energy Consumption Survey. "Where Does My Money Go?" EnergyStar. 2001. http://www.energystar.gov/index.cfm?c=products.pr_pie

Reinert, Bill. "2007 Sustainable Opportunities Summit," speech, March 6, 2007. htpp://pressroom.toyota.com/Releases/View?id=TYT2007 030602653

"Renewable Energy." Dictionary.com. *Dictionary.com Unabridged (v. 1.1)*. Random House, Inc. http://dictionary.reference.com/browse/renewable energy

Resch, Rhone. "Solar Energy Industries Association—Investment Tax Credit Teleconference," November 1, 2007. http://www.seia.org/itc.php

Resch, Rhone. "Solar Energy is Economic Engine for US Economy," news release, January 23, 2008. http://seia.org/solarnews.php?id=158

Resch, Rhone. "Joint Statement Reporting Record Growth of Renewable Energy Sector, Calling for Extension of Renewable Tax Credits for Economic Stimulus," January 22, 2008. http://www.seia.org/solarnews.php?id=156

Rich, Sarah. "Are You Being Greenwashed?" *Businessweek*, March 29, 2007. http://www.businessweek.com/innovate/content/mar2007/id20070329_693675.htm

Richard, Michael Graham. "Important! Why Carbon Sequestration Won't Save Us," *TreeHugger*, July 31, 2006. http://www.treehugger.com/files/2006/07/carbon_sequestration.php

Richardson, Bill. Western Climate Initiative Statement. http://www.westernclimateinitiative.org/ewebeditpro/items/O104F13014.pdf

Richter, Alexander. *United States Geothermal Market Report*, Glitner International Banking, September 2007, 1–37. http://www.glitnirbank.com/services/sustainable-energy

Rodriguez, Nelsy. "Murrieta Resident Seeks to Install Wind Turbine on her Property," *The Californian*, January 19, 2008. http://www.nctimes.com/articles/2008/01/19/news/californian/5_03_991_18-08.prt

Rogers, Jim. "Energy Efficiency: The Fifth Fuel," excerpts of remarks to the Economic Club of Indiana, October 31, 2007. http://www.duke-energy.com/pdfs/Econ-Club-of-Indiana-excerpts-format.pdf

Rogers, James E. "Testimony before House Subcommittee on Energy and Air Quality," Energy and Commerce Committee, March 20, 2007, 1–19. http://energycommerce.house.gov/cmte_mtgs/110-eaq-hrg.032007.Rogers-testimony.pdf

Roth, Bill, "The Green Economic Revolution," *Entrepreneur.com*, April 18, 2008. http://www.entrepreneur.com/management/greencolumnistbill roth/article192980.html

Royal Dutch Shell. "Shell and HR Biopetroleum Build Facility to Grow Algae for Biofuel," news release, November 12, 2007. http://www.shell.com/home/content/media/news_and_library/press_releases/2007/biofuels_cellana_11122007.html

Royal Embassy of Saudi Arabia. "Solar Energy," 2006. http://saudiembassy.net/Country/Energy/EngDetail7.asp

Saffran, Michael. "Senior Project Sheds New Light on the RIT Campus," *RIT News and Events*, May 18, 2007. http://www.rit.edu/~930www/NewsEvenets/2007/May01/t3.html

Savinar, Matt. "Life after the Oil Crash." http://www.lifeaftertheoilcrash.net

Savitz, Andrew. "About the Book," *The Triple Bottom Line*. http://getsustainable.net

Science Daily. "Carbon dioxide sink."http://www.sciencedaily.com/articles/c/carbon_dioxide_sink.htm

Science Daily. "Nanotechnology Now Used in Nearly 500 Everyday Product," May 23, 2007. http://www.sciencedaily.com/releases/2007/05/070523075416.htm

Seafood Choices Alliance. "Trout, Rainbow (farmed)," 2006. http://seafoodchoices.com/smartchoices/species_trout.php

Secter, Bob and Kristen Kridel. "Illinois Lands Clean-Coal Plant," *Chicago Tribune*, December 18, 2007. http://www.chicagotribue.com/news/local/chi-futuregen_webdec19,0,7717400.story

Bibliography

Siegel, Jeff. "Investing in Algae Biodiesel," *Green Chip Stock*, April 24, 2008. http://www.greenchipstocks.com/aqx_p/2575?gclid=CJHcyZ2n7pICFRlP1AodZzZe3g

Siemens, "National Corn-to-Ethanol Research Center Fuels Growth with Siemens," news release, June 29, 2007. http://www2.sea.siemens.com/news/industrial/national-corn-to-ethanol.html

Siemens Solar. "Facts About Solar Energy." http://www.siemensolar.com/facts.html

Skoczek, Marianne. "Center Will Shine New Light on Energy Efficiency Efforts, Entrepreneurship," UC Davis *School News*, June 15, 2006. http://www.gsm.ucdavis.edu/innovator/springsummer2006/eec.pdf

"Smog." Dictionary.com. *The American Heritage® Dictionary of the English Language, Fourth Edition.* Houghton Mifflin Company, 2004. http://dictionary.reference.com/browse/smog

SolarBuzz. "Solar Cell Technologies," 2007. http://www.solarbuzz.com/Technologies.htm

SolarBuzz. "2007 World PV Industry Report Highlights." http://solarbuzz.com/Marketbuzz2008-intro.htm

Solar Development Inc. "Using Solar Energy to Heat Swimming Pools." http://www.solardev.com/SEIA-pools.php

Solar Navigator. "Solar Panels," 1999 and 2008. http://www.solarnavigator.net/solar_panels.htm

South Florida Water Management District. "Reduce Your Water Use, No Excuse!" https://my.sfwmd.gov/portal/page ?_pageid=3074,20103213&_dad=portal&_schema=PORTAL

Southern California Edison. "Edison SmartConnect," information sheet, November 28, 2007, 1–2. http://www .edisonsmartconnect.com.

Southern California Edison. "Edison Smart Connect—Building a Smarter, Cleaner Energy Future with Our Customers," *Southern California Edison Backgrounder*, November 29, 2007. http://www.sce.com/smartconnect.

Southern California Edison. "Edison SmartConnect—Next Generation Metering System," 2008. http://www.sce.com/ PowerandEnvironment/smartconnect/default.htm

Southwest Windpower. "2006 Best Inventions," *Time,* 2006. http:// www.time.com/time/2006/techguide/bestinventions/inventions/ home5.html

Southwest Windpower. "2006 Best of What's New," *Popular Science,* 2006. http://www.popsci.com/popsci/flat/bown/2006/ product_85.html

Southwest Windpower. "Kyoto City Government Chooses Small Wind/Solar Systems for Its 121 Schools' Environmental Educa-tion," news release, December 2006. http://www.windenergy .com/globalinstalls/global_installs_kyoto.htm

Southwest Windpower. "Navajo Project," news release. http:// www.windenergy.com/globalinstalls/navajo_project.htm

Southwest Windpower. "Southwest Windpower," 2007. http:// www.windenergy.com/index_wind.htm

Bibliography

Southwest Windpower. "Southwest Windpower Raises $8 Million in Growth Capital," news release, April 7, 2006. http://www .windenergy.com/news/new6a.htm

Spadaro, Joseph V., Lucille Langlois, and Bruce Hamilton. "Greenhouse Gas Emissions of Electricity Generation Chains: Assessing the Difference," *IAEA Bulletin,* 42/2/2000. http:// www.iaea.org/Publications/Magazines/Bulletin/Bull422/ article4.pdf

Specter, Bob, and Kristen Kridel. "Illinois lands clean-coal plant," *Chicago Tribune* online, December 18, 2007. http://www .chicago tribune.com

Sterling Planet. "3 Steps to Carbon Neutrality," pamphlet, 1–2. http://www.sterlingplanet.com/upload/File/Sterling %20Planet%203%20Steps%20to%20Carbon%20Neutrality %20Fact%20 Sheet.pdf

SunPower. "About Us." http://www.sunpowercorp.com/about -us.aspx.

SunPower. "Buy Electricity Only." http://www.sunpowercorp.com/ For-Businesses/How-To-Buy/Buy-Electricity.aspx

SunPower. "Energy Efficiency." http://www.sunpowercorp.com/ Products-and-Services/Energy-Efficiency.aspx

SunPower. "Nellis Air Force Base Builds Largest Solar Pho- tovoltaic Power Plant in North America with SunPower," case study. http://sunpowercorp.com/For-Power-Plants/ Case-Studies.aspx

SunPower. "Performance Monitoring." http://www.sunpowercorp .com/Product-and-Services/Performance-Monitoring.aspx.

Sustainable Agriculture Research and Education. "The Patrick Madden Award for Sustainable Agriculture." http://www.sare .org/coreinfo/madden2006.htm

Sustainable Executive Order 13423. *Federal Register*, January 26, 2007. http://ofee.gov/eo/EO_13423.pdf

Sustainable Forestry Initiative. "SFI Standard." http://sfiprogram .org/standard.cfm

Sustainable Forestry Initiative. "SFI Products and Forests. Certified Forests Search." http://www.certifiedwoodsearch.org/ sfiprogram

SustainLane. "Portland: A Role Model for the Nation," 2006. http://www.sustainlane.com/us-city-rankings/portland.jsp

Taylor, James M., and Joseph L. Bast. "What Is Commonsense Environmentalism," *Environment Issue Suite*, April 16, 2007. http://www.heartland.org/Article.cfm?artID=10488

TerraChoice. "The Six Sins of Greenwashing™—A Study of Environmental Claims in North American Consumer Markets," *Terrachoice.com*, November 2007, 1–12. http://www.terrachoice .com

TerraCycle. "Cookie Wrapper Brigade." http://www.terracylce.net/ ob/ob.htm

TerraCycle. "The TerraCycle Story," 2007. http://terracycle.net/ story.htm

Thompson, Jon F. "Toyota and Sustainable Mobility," Lexus weblog, February 11, 2008. http://blog.lexus.com/being_green/ index.html

Bibliography

Thomson Reuters. "DOE and National Laboratories Project Targets Commercial Viability for Enhanced Geothermal," February 14, 2008. http://www.reuters.com/article/pressRelease/idUS245936+14-Feb-2008+PRN20080214

Travel Portland. "The Deals—Green/Sustainable." http://www.travelportland.com/deals/green_sustainable.html

Toyota. "How Hybrid Vehicles Work," news release, October 1, 2007. http://pressroom.toyota.com/Releases/View?id-TYT2007100288452

Toyota. "Fairbanks-to-Vancouver Along the Alaska Highway," news release, November 14, 2007. http://pressroom.toyota.com/Releases/View?id=TYT2007111310632

Toyota. "Toyota 1/X Concept Makes North American Debut at 2008 Chicago Auto Show," news release, February 6, 2008 http://pressroom.toyota.com/Releases/View?id=TYT200802000 168848

Toyota. "Toyota Environmental Update," news release, February 13, 2008. http://pressroom.toyota.com/Releases/View?id=TYT2008021335516

Toyota. "Toyota Promotes Plug-in Hybrid Development at the 2008 North American International Auto Show," news release, January 14, 2008. http://pressroom.toyota.com/Releases/View?id+TYT2008011405523

Toyota. "Toyota to Display CNG-Powered Camry Hybrid Concept at 2008 Los Angeles Auto Show," news release, September 24, 2008. http://pressroom.toyota.com/Releases/View?id=TYT2008092370279

Toyota. "Toyota Vehicles Earn Top Ratings in 2008 Environmental Vehicle Guide," Environmental Update, forty-ninth issue, April 2008.http://www.toyota.com/about/news/environment/2008/01/04-1-2008AprEnvUpdate-49.html

Unimodal Inc. (SkyTran). "The SkyTran Experience," 2004. http://www.unimodal.net

University of California, Davis, Energy Efficiency Center. "Welcome to the Energy Efficiency Center." http://eec.ucdavis.edu

University of California, Davis, "New Technology Turns Food Leftovers Into Electricity, Vehicle Fuels," October 24, 2006. http://www.news.ucdavis.edu/search/news_detail.lasso ?id=7915

United Nations Framework Convention on Climate Change, "Welcome to the United Nations Climate Change Conference in Bali." http://unfccc.int/meetings/cop_13/items/4094.php.

United Solar Ovonic. "Building Integrated System Solutions." http://www.uni-solar.com/interior.asp?id=83

United Solar Ovonic. "Solar Laminates." http://www.uni-solar.com/interior.asp?id=102

United Solar Ovonic. "United Solar Ovonic Announces Agreement with SunEdison to Supply Up to 17 MW of Photovoltaic Laminates," news release, December 10, 2007. http://www.uni-solar.com/newsdetail.asp?id=219

United States Census Bureau. "Americans Spend More than 100 Hours Commuting to Work Each Year, Census Bureau

Reports," news release, March 30, 2005. http://www
.census2010.gov/Press-Release/www/releases/archives/
american_community_survey_acs/004489.html

United States Climate Action Partnership, "Joint Statement
of the United States Climate Action Partnership," news
release, January 19, 2007. http://us-cap.org/media/release
_USCAPStatement011907.pdf

United States Climate Action Partnership. "A Call For Action,"
January 22, 2007, 2–12. http://us-cap.org/USCAPCallForAction
.pdf

United States Department of Labor—Bureau of Labor Statistics.
Federal Government, Excluding the Postal Service, March 12,
2008. http://www.bls.gov/oco/cg/cgs041.htm

United States Fish and Wildlife Service. "Homeowners Guide to
Protecting Frogs—Lawn and Garden Care." http://www.fws
.gov/contaminants/documents/homeowners_guide_frogs.pdf

UNFCCC. "Article 2: Objective," United Nations Framework
Convention on Climate Change. http://unfccc.int/essential
_back ground/convention/background/items/1353.php

UPC Wind. "UPC Wind Obtains Tax Equity Financing for Its
2008 New York Portfolio," news release, February, 4, 2008.
http://www.upcwind.com/aboutUPC/news.cfm

USAID. "USAID Support for the Wind Power Industry—The
Clipper Example," August 30, 2007. http://www.usaid.gov/our
_work/environment/climate/pub_outreach/story_mexico.html

USDA. "Community Supported Agriculture," National Agricultural Library, May 29, 2008. http://www.nal.usda.gov/afsic/pubs/csa/csa.shtml

U.S. Green Building Council. "238 New Developments Nationwide Join Pioneering LEED for Neighborhood Development Pilot..." August 8, 2007. http://www.usgbc.org/News/PressReleaseDetails.aspx?ID=3304

Velasquez-Manoff, Moises. "Before Regulation Hits, a Battle Over How to Build New U.S. Coal Plants," *The Christian Science Monitor*, February 22, 2007. http://www.scmonitor.com/2007/0222/p13s01-sten.htm

von Hippel, Frank N. "Nuclear Fuel Recycling: More Trouble Than It's Worth," *Scientific American*, April 28, 2008. http://www.sciam.com/article.cfm?id=rethinking-nuclear-fuel-recycling

Wagoner, Rick. "Electric Avenue—the Convergence of Electronic and Automotive Technologies," speech, January 8, 2008. http://media.gm.com/us/gm/en/news/speeches/index.html

Wagoner, Rick. "Re-invention in a Rapidly Changing World," speech, May 8, 2007. http://media.gm.com/us/gm/en/news/speeches/index.html

Wagoner, Rick. "Remarks at the GM Bio-Fuels Press Conference," speech, January 13, 2008. http://media.gm.com/us/gm/en/news/speeches/index.html

Wald, Mathew L. "EPA Is Prodded to Require Cuts in Airliner Emissions," *The New York Times*, December 6, 2007. http://www.nytimes.com/2007/12/06/washington/06planes.html?partner=rssnyt&emc=rss

Bibliography

Walsh, Bryan. "Going from Bacteria to Gasoline," *Time*, December 7, 2007. htpp://www.time.com/time/magazine/article/0,9171,1692404,00.html

Werner, Tom. "SEIA—ITC Press Teleconference," November 1, 2007. http://www.seia.org/itc.php

Weyerhaeuser. "Chevron and Weyerhaeuser Create Biofuels Alliance," news release, April 12, 2007. http://www.weyerhaeuser.com/Company/Media/NewsReleases

Weyerhaeuser. "Chevron and Weyerhaeuser for Biofuels Joint Venture," news release, February 29, 2009. http://www.weyerhaeuser.com/Company/Media/NewsReleases

Weyerhaeuser. "Sustainable Systems," September 17, 2007. http://www.weyerhaeuser.com/Sustainability/Systems

Wind & Hydropower Technologies Program. "Wind Power Pioneer Interview: Jim Dehlsen, Clipper Windpower," October 1, 2003. http://www.eere.energy.gov/windandhydro/windpoweringamerica/filter_detail.asp?itemid=683

Wind Energy News. "AeroCity's Urban Wind Power Turbine," October 17, 2008. http://www.windenergynews.com/content/view/1432/43

Woody, Todd. "Clock Ticking on Crucial Solar Investment Tax Credit," enviroblog, January 22, 2008. http://greenwombat.blogs.fortune.cnn.com/2008/01/22/clock-ticking-on-crucial-solar-investment-tax-credit

World Coal Institute. "Coal Facts 2007," October 2007. http://www.worldcoal.org/pages/content/index.asp?PageID=188

World Nuclear Association. "US Nuclear Power Industry," October 2007. http://www.world-nuclear.org/info/inf41 .html#reprocessing

Worldwatch Institute. "Local Food—Did You Know?" http://www .worldwatch.org/node/4132

Wynn, Will. "Small Businesses at the Forefront of the Green Revolution: What More Needs to Be Done to Keep Them Here?" Testimony for House Committee on Small Business, July 11, 2007. http://www.cityofaustin.org/council/downloads/ mw_plugin_testi mony.pdf

Websites:

American Automobile Association (AAA). http://www .fuelguagereport.com

Amtrak. http://www.amtrak.com

AeroCity LLC. http://www.directglobalpower.com

Aircuity Inc. http://aircuity.com

American Electric Power (AEP). http://www.aep.com

American Petroleum Institute (API). http://www.api.org

American Wind Energy Association (AWEA). http://awea.org

Austin Energy. http://www.austinenergy.com

BigBellySolar. http://bigbellysolar.com

Bioplastics24.com. http://www.bioplastics24.com

Bibliography

BP. http://www.bp.com

BP Solar. http://www.bpsolar.com

BrightSource Energy. http://www.brightsourceenergy.com

California Institute for Telecommunications and Information
Technology (CalIT2). http://www.calit2.net

California Energy Commission (CEC). http://www.energy.ca.gov

California Hydrogen Highway. http://www.hydrogenhighway
.ca.gov

Calpine. http://www.geysers.com

Carbon Tax Center (CTC). http://www.carbontax.org

Cape Wind. http://www.capewind.org

Center for Small Business and the Environment (CSBE).
http://www.aboutcsbe.org

Chicago Climate Exchange (CCX). http://www.chicagoclimatex
.com

City of Austin. http://www.ci.austin.tx.us

City of Miami. http://www.miamigov.com

City of Portland. http://www.portlandonline.com

Clean Energy Alliance. http://www.cleanenergyalliance.com

Clean Energy Incubator (CEI). http://www.cleanenergyincubator
.org

Clipper Windpower. http://www.clipperwind.com

Congress for the New Urbanism (CNU). http://www.cnu.org

Constructive Technologies Group (CTG). http://www.ctg-net.com

Cornell University Northeast Sun Grant Initiative, Institute of
Excellence. http://www.nesungrant.cornell.edu/cals/sungrant

Coskata. http://www.coskataenergy.com

Defense Advanced Research Projects Agency (DARPA), Small
Business Innovation Research (SBIR) Program and Small
Business Technology Transfer (STTR) Program. http://www
.darpa.mil/sbir

Database of State Incentives for Renewable Energy (DSIRE).
http://www.dsireusa.org

Department of Energy (DOE). http://www.energy.gov

DOE, Energy Information Administration (EIA). http://www.eia
.doe.gov

DOE, Labs and Technology Centers. http://www.doe.gov/
organization/labs-techcenters.htm

DOE, Office of Civilian Radioactive Waste Management
(OCRWM). http://www.ocrwm.doe.gov

DOE, Office of Energy Efficiency and Renewable Energy (EERE).
http://www.eere.energy.gov

Bibliography

DOE, Office of Nuclear Energy, Global Nuclear Energy Partnership (GNEP). http://www.gnep.energy.gov

Duke Energy. http://www.duke-energy.com

Duke University Division of Earth & Ocean Sciences. http://www.nicholas.duke.edu/eos/

Eastern Electric Vehicle Club (EEVC). http://eevc.info

Edison International. http://www.edison.com

Encyclopedia of Alternative Energy and Sustainable Living http://www.daviddarling.info/encyclopedia/AEmain.html

Environmental Defense Fund (EDF). http://www.edf.org

Environmental Protection Agency (EPA). http://www.epa.gov

EnergyStar. http://www.energystar.gov

Energy Trust of Oregon. http://www.energytrust.org

Extengine. http://www.extengine.com

Federal Energy Regulatory Commission (FERC). http://www.ferc.gov

Fleets and Fuels. http://fleetsandfuels.com

Flex Your Power. http://www.flexyourpower.org

FutureGen Alliance. http://www.futuregenalliance.org

Georgia Institute of Technology, Institute of Paper Science and Technology (IPST). http://www.ipst.gatech.edu

General Atomics (GA). http://www.ga.com

General Electric (GE). http://www.ge.com

GE Ecomagination. http://ge.ecomagination.com/site

GE Energy. http://www.gepower.com

General Motors (GM). http://www.gm.com

Generation IV International Forum (GIF). http://www.gen-4.org

Geothermal Energy Association (GEA). http://www.geo-energy.org

Geothermal Heat Pump Consortium (GHPC). http://geoexchange
.us

Greenbuild Expo. http://www.greenbuildexpo.org

Green Chip Stocks. http://www.greenchipstocks.com

GreenFuel Technologies Corporation. http://www.greenfuelonline
.com

Green Seal. http://www.greenseal.org

HR BioPetroleum (HRBP). http://www.hrbiopetroleum.com

Hydrogen Filling Stations Worldwide. http://www.h2stations.org

Innovative Transportation Technologies. http://faculty.washington
.edu/jbs/itrans

Institute of Self Reliance. http://www.ilsr.org

International Atomic Energy Agency (IAEA). http://www.iaea.org

Bibliography

International Council for Local Environmental Initiatives (ICLEI). http://www.iclei.org

International Ground Source Heat Pump Association (IGSHPA). http://www.igshpa.okstate.edu

International Thermonuclear Energy Reactor (ITER). http://www.iter.org

Iowa Heat Pump Association (IHPA). http://www.iaheatpump.org

Konarka. http://konarka.com

Lawrence Livermore National Laboratory (LLNL). https://www.llnl.gov

LS9. http://www.ls9.com

Marine Stewardship Council (MSC). http://www.msc.org

Miami 21. http://www.miami21.org

MegaRail Transportation Systems Inc. http://www.megarail.com

Metabolix. http://www.metabolix.com

National Biodiesel Board. http://www.biodiesel.org

National Corn-to-Ethanol Research Center (NCERC). http://www.ethanolresearch.com

National Energy Laboratory of Hawaii Authority (NELHA). http://www.nelha.org

National Energy Technology Laboratory (NETL). http://www.ntel.doe.gov

National Ethanol Vehicle Coalition (NEVC). http://www.e85fuel
.com

National Fuel Cell Research Center (NFCRC). http://www.nfcrc.uci
.edu

National Hydropower Association. http://www.hydro.org

National Pollution Prevention Roundtable (NPPR). http://www
.p2.org

National Renewable Energy Laboratory (NREL). http://www.nrel
.gov

Natural Resource Defense Council (NRDC). http://www.nrdc.org

Nature Conservancy. http://www.nature.org

New York State Energy Research and Development Authority
(NSERDA). http://www.nyserda.org

Northeast Sustainable Energy Association (NESEA). http:www
.nesea.org

Oak Ridge National Laboratory. http://www.ornl.gov

Ocean Renewable Power Company, LLC (ORPC).
http://oceanrenewablepower.com

Orangeline Development Authority. http://www.orangeline
.calmag lev.org

Oregon State University Wallace Energy Systems and Renewables Facility (WESRF). http://eecs.oregonstate.edu/wesrf

Pelamis Wave Power. http://www.pelamiswave.com

Bibliography

People History, The. http://www.thepeoplehistory.com

Plug Power. http://www.plugpower.com/technology/hydrogen.cfm

Prism Solar Technologies, Inc. http://www.prismsolar.com

Puna Geothermal Venture (PGV). http://www
.punageothermalventure.com

Santa Clara University Solar Decathlon Team. http://www
.scusolar.org

Seafood Choices Alliance. http://www.seafoodchoices.com

Solar Decathlon. http://www.solardecathlon.org

Solar Energy Industries Association (SEIA). http://www.seia.org

SunPower Corporation. http://www.sunpowercorp.com

Southern California Edison (SCE). http://www.sce.com

SCE, San Onofre Nuclear Generating Station (SONGS).
http://www.sce.com/PowerandEnvironment/PowerGeneration/
SanOnofreNuclearGeneratingStation

Southwest Windpower (Skystream 3.7). http://
www.skystreamenergy.com

State Energy Conservation Office (SECO). http://www.seco.cpa
.state.tx.us

Sustainable Forestry Initiative (SFI). http://www.aboutsfi.org

SustainLane. http://www.sustainlane.com

Technische Universitat Darmstadt Solar Decathlon Team.
http://www.solardecathlon.de

TerraCycle. http://terracycle.net

TerraPass. http://www.terrapass.com

Toyota. http://www.toyota.com

Travel Portland. http://www.travelportland.com

Troutlodge Inc. http://www.troutlodge.com

Unimodal Inc. (SkyTran). http://www.unimodal.net

United Solar Ovonic. http://www.uni-solar.com

United States Climate Action Partnership (USCAP). http://www
.us-cap.org

United States Department of Agriculture (USDA), Cooperative
State Research, Education and Extension Service (CSREES).
http://www.csrees.usda.gov

USDA, Sustainable Agriculture Research and Education (SARE).
http://www.sare.org

US Department of Commerce, National Oceanic and Atmospheric
Administration. http://www.noaa.gov

US Department of Interior, Minerals Management Services
(MMS). http://www.mms.gov/mmshome.htm

US Department of Transportation (DOT). http://www.dot.gov

Bibliography

US Green Building Council (USGBC). http://www.usgbc.org

US Nuclear Regulatory Commision (NRC). http://www.nrc.gov

University of California, Davis, Center for Entrepreneurship. http://entrepreneurship.ucdavis.edu

University of California, Davis, Energy Efficiency Center (EEC). http://eec.ucdavis.edu

University of California, Davis, Green Technology Entrepreneurship Academy (GreenTEA). http://entrepreneurship.ucdavis.edu/green

University of California, Davis, Institute of Transportation Studies (ITS). http://www.its.usdavis.edu

University of California, Irvine, Advanced Power and Energy Program (APEP). http://www.apep.uci.edu

University of California, Irvine, National Fuel Cell Research Center (NFCRC). http://www.nfcrc.uci.edu

University of Illinois at Urbana-Champaign, Agroecology and Sustainable Agriculture Program (ASAP). http://asap.sustainability.uiuc.edu

University of Maryland Solar Decathlon Team. http://www.solarteam.org

University of Southern California Marshall School of Business Lloyd Greif Center for Entrepreneurship. http://marshall.usc.edu/greif

University of Southern California Stevens Institute for Innovation. http://stevens.usc.edu

University of Southern California Wrigley Institute for Environmental Studies. http://wrigley.usc.edu

Verdant Power. http://verdantpower.com

Waste Management and Research Center (WMRC). http://www.wmrc.uiuc.edu

WaterSaver Technologies. http://www.watersavertech.com

Wave Dragon. http://www.wavedragon.co.uk

Wavegen. http://www.wavegen.co.uk

West Texas A & M University, Alternative Energy Institute. http://www.windenergy.org

Weyerhaeuser. http://www.weyerhaeuser.com

White House, The. http://www.whitehouse.gov

Wikipedia. http://en.wikipedia.org

Zero Emission Vehicle-Network Enabled Transport (ZEVNET). http://www.zevnet.org

Index

Index

Index

Index

Index

Index